The History of Urban Planning and Cities

The cover art is a map of the ancient city of Rome. No city had a larger impact on the western world and political systems than ancient Rome. We owe our current Senate, our transportation system network, our civil engineering projects, our athletics systems, our military system to the Romans who were the most efficient rulers of the ancient world with influence in most of the European and Mediterranean Theaters of the world. If you look closely you can see the Coliseum at the center of the city. A special thanks to Yahoo and other sources for all the maps in this book.

Forward

This book was written to help anyone understand the development of the cities

they visit as well as instruct urban planning students on some of the historical

development aspects of major cities in the world. I have not yet visited every city

in this compendium, but it is on my priority list. Hopefully, the young urban

planning students will have a desire to manage these cities and see the beautiful

architecture and civil engineering projects that each city has undertaken in

building the modern world we live in. It is hoped that this book can augment an

understanding of ancient cities, archeology, macro-architecture, and civil

engineering. Any errors in the content are strictly mine and your comments are

greatly appreciated. I have used the broadest definition of software in creation of

this book, that is, Congress now considers all forms of media to be software. I

have also used my experience with federal, state, and local governments and

councils in this book as a taxpaying citizen, lobbyist, civil servant, and Chief

Voting Judge. I hope you enjoy this book. Happy hunting!

Donald J.G. Chiarella, Ph.D., MSTM, BA, CISM, CDRM

Table of Contents

List of Figures

1. Mesopotamia

Mesopotamia is known as the cradle of mankind. Akkadia was the oldest of the civilizations in Mesopotamia from between 2350-2230 B.C. with cities such as Akkad, Ur, and Erich[1].

Ancient Babylon is in the middle of Mesopotamia. Ancient Babylon stories permeate the bible. The story of the Tower of Bable took place in Babylon. The Euphrates River valley runs through Babylonia near current day Baghdad and Iraq. Ancient Babylonia was a powerful nation between 1728-1686 B.C. and 625-538 B.C.[2] It was King Nebuchadnezzar who built the empire and defeated Jerusalem (586 B.C.) and thus captured the Jews. The Hanging Gardens of Babylon are one of the seven wonders of the ancient world. Persia then conquered Babylon and freed the Jews. This is also documented in the Old Testament.

Today we call this area the middle east. It included Palestine, Iran and parts of Asia Minor. It makes sense that on Sundays we study the ways of the bible and we learn about life in the cities of the middle east. Surely, life was harder on those people than it is on modern man with all his amenities.

---

[1] 2005 Time Almanac, pp 669.
[2] Ibid, pp 669

Below the southern edge of Mesopotamia there was the Nile Valley and Egypt. The Egyptians are considered some of the most advanced people of earth at that time from 2850-715 B.C. with major cities such as Thebes, Memphis, and Tanis.

Syria was another country in Mesopotamia[3]. The Assyrian culture lived between 1800-899 B.C in Mesopotamia and the major cities were Assur, Nineveh, and Calah. The Sumerians inhabited Mesopotamia as well from 3200-2360 B.C. in the cities of Ur and Nippur.

Buddha appeared in India 563 years before Jesus walked the earth. Had they met, surely they would have shaken hands and become friends. Today, Buddhist countries include Japan, China, Vietnam, Laos, Cambodia, and Thailand. It is clear that Mesopotamia influenced the other cultures to the east as people traveled between them. We know that Mesopotamia influenced the western cultures of Greece and the Etruscans as well as the Hittites (1640-1200 B.C.).

Man has always located civilizations near bodies of water for survival reasons. From Mesopotamia cultures began appearing in all directions of the compass over the years always near a water supply.

Figure 1 shows Africa, the Mediterranean, and Mesopotamia current day.

[3] Ibid., pp 669

Figure 1. Africa and Mediterranean (Ancient Mesopotamia)

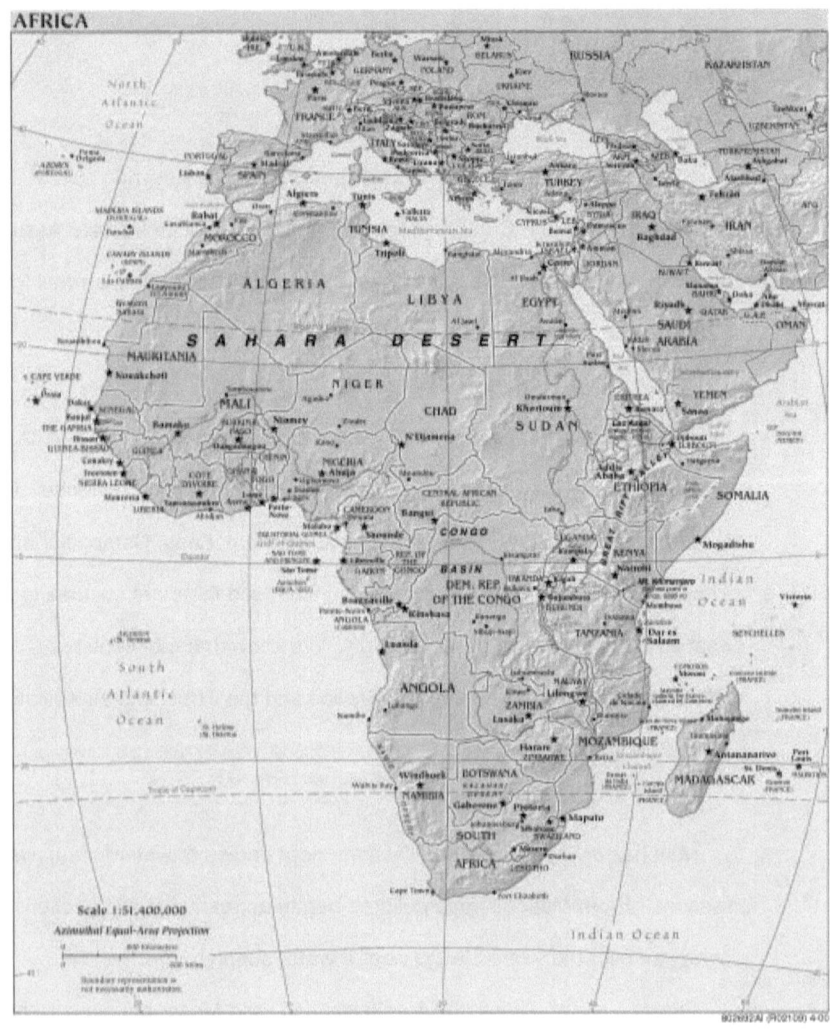

2. Ancient Egyptian and Mayan Civilizations

These two ancient cultures are important because they were so well developed architecturally that they left behind major cities at Giza and Perengue[4] where they built their pyramids and structures in the form of ancient cities.

Ancient Egypt was important to the development of the middle east because they were the power of the time and had excellent engineers and professionals in their society. The Romans visited Egypt and took ideas back to Rome for building and architectural changes. When Mark Anthony and Cleopatra were in love, many ideas exchanged between the two cultures. The Egyptians were led in the building of the city of Giza near Goshen by Moses and this is chronicled in the Holy Bible in Exodus. Moses was the perfect master builder among his many other skills. He built the city with slave labor according to Jewish Traditions. The history channel[5] stated that more Egyptians were used on the pyramids than previously believed because the Pharoe Ramsees wanted to employ his own trusted people. The pyramids were each so large that we are still puzzled at the exact techniques of ramps and inclines used to built them and lift the 20 tons blocks up the side of each pyramid over a 20 year time frame. The barges brought the cut blocks from far away down the Nile and were unloaded in Giza. The Nile was considered a life source to the Egyptians and played a significant role in

---

[4] History Channel Special "Lost Worlds: Perengue" aired in April 2005
[5] History Channel Special on "Ancient Pyramids of Egypt" aired in 2004

everyday life. Teams were built, housed, and fed, to do all the work over the 20 years. It is clear the monumental task required both extensive time and labor like the world has never seen again. This may be why the pyramids still stand today as one of the ancient wonders of the world. The city was built with 16 large lions at the gate and large obelisks paying homage to the Pharoe. The largest pyramid of those built has a hidden tunnel to the inner chamber room where the Pharoe was buried. He was buried with possessions from his life in a sarcophagus after embalming. Today the city is a tourist attraction for those willing to venture there to see the great pyramids.

It is clear that both Giza at ancient Egypt, built by Moses and Perengue used similar building techniques, but the experts can not figure how they were so similar in structure when they were so many miles apart. Perengue is in Central America and was built around 548 AD. It was built by a king who wanted the city to flourish with more than 100,000 inhabitants. Archeologists have uncovered all the dwellings and mapped the inner CBD which was mostly temples and pyramids. The residences were in what would have been the local suburbs. It is totally covered in moss and vines today but tourists may visit. Scientists discovered the tomb of the king was built like the ancient Egyptian tombs. The city flourished for 800 years and then the people just vanished. The Mayans also build sports arenas, sun temples to the Gods, and a tall 7 story tower in the downtown area to tell if trouble approached. The tomb of the king had tunnels to the chamber vault where he was buried in a large vault very similar the Egyptian Pharos at Giza. The were no tall obelisks at Perengue. The Mayans used simple geometry as an engineering guideline. They used a circle shaped like a lotus blossom and the square to tell angles of all the buildings. Every building in Perengue uses this ancient geometry.

3. African Tribes

African tribes are many in the nations of Africa. Egypt is among these

northern African tribes. Egypt was considered one of the most advanced

cultures on earth in the ancient world by Rome and Greece who sent doctors

there to study Egyptian medicine. Ethiopia is another ancient culture who

once warred with Egypt when Moses was a prince in the Royal family of

Egypt before the times of Exodus in the Bible. Kenya is another nation with

most recent famous Olympians in long distance running. America was

introduced to African tribalism during the miniseries "Roots"[6] which was very

popular. It was the story of Alex Haley's quest for his family ties from Africa to

the deep south. Now we celebrate February as Historic African American

month along with Dr. Martin Luther King's birthday in January. Both are fitting

for a strong people of such endurance and courage under slavery in America[7]

[8] and the world before Emancipation by President Lincoln. Today, Americans

live side by side and work side by side and die side by side on the battlefields

for freedom. Hopefully, our children of all colors do not take this freedom for

granted.

The Scottish were instrumental in trying to help Kenya build a new

government. The British have interests in South Africa especially under

---

[6] ABC Mini-series entitled "Roots".
[7] Book, The Life of Frederick Douglass.

[8] Book, Uncle Tom's Cabin, Harriet B. Stowe.

Apartheid and even today (diamond mines and other natural resources).
America has interests in seeing all of Africa live free from dictatorship in
democracies. The aides epidemic runs rampant on the continent with some
even believing that they will not be infected if they have sex with a virgin.
Americas churches have a conscience and are trying to help countries like
Liberia (civil war torn) and Rowanda[9]. I posit that most White or Asian people
do not understand what it means to be an African American today. In
America, most Whites and Asians are very high income producers compared
to African Americans on the whole. This is primarily because we use the
White man's education system to judge everyone by. I prefer to call it the
Christian education system. I know different. Many well intended Christians
insult our African American brothers every day by assuming they always need
a hand out instead of a hand up. The simple acts of kindness we show
everyone must include our African American brothers and sisters. This is a
Christian value in my family, a middle class white American family, and a
value of judging people for who they are not what color they are. The tribes
of Africa fought to keep their sons and daughters from slavery. Some sold
them into slavery as a last resort. Can anyone believe they did this of free
will? America has lived through Somalia and other war torn efforts such as
the Sudan and we do not do enough every day to help these people obtain
freedoms and human rights they deserve. The only people I see trying to do
something are President Carter through his Carter Center, President Bush

---

[9] Movie based on real story entitled "Hotel Rowanda".

(Sr.), President Clinton both through the Tsunami efforts (and asked by younger Bush).

In the early 1990's some British and American rock stars also held concerts and rallies for the benefit of worldwide aides victims. We only hope they received as much aide as possible. Are we a nation so who thinks we do not have an obligation to these poor countries in favor of foreign politics and conquests that favor the resource rich countries alone? That game has been played by the British Empire and others (Japan) who long ago depleted their natural mineral and petrol resources. Fortunately, we are a nation with a heart and are the Defacto caretaker of the free world. We are a compassionate country even before that term became political.

Some good books on the struggles of black people in America include Uncle Tom's Cabin, Black Like Me, Native Son, and The Narrative of the Life Of Frederick Douglas. Then we need to reread the speeches of reverend Martin Luther King, especially the "I have A Dream Speech". I recently found out his real first name was Michael as was his father's. His father wanted him to be a pastor so he changed the name o match the founder of Protestantism. It does not diminish what he accomplished in his short 39 year lifetime through peaceful demonstration means. Were I a young African American trying to understand white America, I would watch the "West Side Story", "The Greatest Story Ever Told", and all the Cary Grant movies. These cover three important areas of white life in America, the White gang groups in cities in the 1950's - poor whites, The Christian right white groups of rural and urban America, and the Upper Class white parents of yuppies. Cary Grant never did a movie without a nice suit on so he appealed to most business and executive types of the day of all colors.

Interviews and some research led me to some interesting conclusions about modern Africa. Many of the nation states have Kings and Royal families that where begun after Christ was alive (AD). These same nation states are now evolving into democracies who may or may not want outside help from America. To assume they always want outside help is wrong and misunderstands the variety of African tribal subcultures. Some of the more affluent tribes are living good lives by African standards. We in the west see movies like Zulu and Doctari and form our opinions. Africa has many people from all over the world like India, Europe, Asia, and America living there. To accept them as a multicultural country is appropriate by the rest of the world. I need to visit Africa, but I ask what can I do to help their long term suffering, so I have studied them from afar since childhood preferring to study the Egyptians and ignoring the poorest nations. To meet a Christian who understands that Moses was an African man (from Egypt) is to say the least to meet someone who understands where we all come from historically near Mesopotamia in the Fertile Nile region. The Catholic church has made graet advances in Africa and Latin America in terms of new membership.

4. Native American Tribes

Every time I visit the Midwest and west I am reminded of what I knew when I was a child living in Topeka, Kansas about the American Indian. They got the worst deal from the Europeans than anyone in history, except African Americans. The word Kansas is named after a tribe of local Native Americans the Kansa tribe. Topeka is an Indian word. Many of our cities are named after their Indian names such as Milwaukee, Kansas City, Sioux City, and many others. We just never think about it, especially on the east coast. It's sad that the best we can give them are places like the "badlands' in South Dakota that are not fit to live in according to our luxurious modern lifestyles. Arizona is another place we gave the Navajo Indians. They gave us the rest of America. We owe a great deal to the Native Americans and making them the custodians of gambling casinos is a despicable way of showing our affection towards them. They have much they can teach us about their way of life and the spiritual world they believe in. They gave us the Eagle as our national symbol. They gave us many military code names and indeed helped win World War II with the code breakers speaking Navajo and confusing the Japanese. More Native Americans have been in the Marines Corps by percentages of their population than any other group of people. The white man was good at recognizing the warrior abilities of the Native Americans and our high schools, colleges, and team names reflect this honor. However,

24

team names like the Redskins truly are derogatory considering this was a rotten name the white man created for Native Americans. Names such as Warriors, Black hawks, Arrowhead Stadium, Chiefs all honor our Native American brothers and sisters. Many of them became Christians. When was the last time you heard of a white person becoming Native American? (exception given to Kevin Costner in the "Dances with Wolves" movie). Our history books only tell the white story of history and the Native American knows this. We need to show more sensitivity, grace, and tolerance in our writings on other cultures.

With over 1,056,457 Native Americans in the United States[10] they should receive better treatment. The tribe with the largest number of people is the Cherokee tribe with 729,533 people. Second is the Navajo with 298,197. Third is the Choctaw with 158,774 people. The Sioux have 153,360 people. Then fifth is the Chippewa with 149,669 people. The largest Federal Indian reservations are in Arizona by land mass and then South Dakota. Third is Montana, then Utah, then New Mexico. Native Americans made up 2.8% of the total population in America in the 2000 census data.

---

[10] 2005 Time Almanac, Page 378.

Figure 2.   Native American Reservation Populations

Tribe	Population	Tribe	Population
Apache	96,833	Navajo	298,197
Blackfeet	85,750	Pueblo	74,085
Cherokee	729,533	Seminole	27,431
Cheyenne	18,204	Sioux	153,360
Chickasaw	38,351	Ute	10,385
Chippewa	149,669	Eskimo	54,761
Comanche	19,376	Not Specified	195,902
Creek	71,310	Choctaw	158,774
Iroquois	80,822	Total	4,119,301

Source: Time Almanac 2005 p. 378

5. European Cities

The European cities all have a similar look and feel of aged dignity. They were mostly here long before any of the American and Canadian cities and are in very dire need of sand blasting of the older buildings and cleaning up of the cities in general. This is what makes them distinctive and many architects and engineers travel to European cities to study how they were built. In fact, the architects and builders in Scotland visited ancient ruins in Rome and Greece to gain insights into early western architecture and engineering. These things were done very early in the Roman empire and Greek world. It follows logically, that Americans would visit all cities in Europe for a better understanding of historic buildings and methods of the past. For the most part the cities on the east coast of the United States that were built in the 1700's and 1800's used European methods and styles in the city development. The farther westward in America you travel, the more modern the city development you will encounter simply because this is the way America was purchased and developed by the US Government. It is very refreshing to see the new architecture and engineering in the western United States and gives hope that we may just be improving our methods of building cities. This does not mean that traditional building types do not have their place in modern society just that we can mix styles as we create new and old ones along side each other in the new cities. Surely, the new cities on other planets and space colonies will have their own distinctive types different from any earthly city without the same environmental concerns and greater need for environmental engineering and protection in non-earth atmospheres. Europe was the cradle of western society and we need to relish in the style they used in the middle ages such as castles, villages, forts, and river side cities.   The cities in Europe expect to communicate with other countries via roadways and rail systems unlike in America where we rarely think like this. They also are better prepared to speak more languages than the United States when they travel to other cities and countries nearby. This diversity is lost on most Americans and American politicians who think English should be the only language when, in fact, the pope always speaks at least 10 languages to communicate with his global flocks. That is reality of the world, not America closed mindedness. In fact, the pope needs to learn more Asian languages to open up his church to those people. Vatican City is the smallest important nation in the world should take the lead on international relations. It has done this in smaller third world countries effectively where the United States has failed diplomatically. Certainly other churches designed in the various cities around the world would emulate St Peter's Basilica due to it's wonderful architecture and design if they could afford the cost.

6. The Baroque Style City

The Baroque style of city was exemplified in Paris in the 1600-1700s. The style was brought to America and practiced in city planning of Williamsburg, old town Philadelphia, and Washington DC. Pierre L' Enfant was the man responsible for designing Washington DC and Paris so he had a unique perspective on the designs of both these cities. Williamsburg is navigated by foot in the old part of the town with period home and taverns still in excellent shape from original 1600's construction.

The early settlers were mostly English in Williamsburg. The French – British connection would have gone back to 1066 with the Norman conquest and William the Conqueror taking the throne of England at the Battle of Hastings in England. It would have been natural for them to share cultural and civilization construction and urban architecture being so close to each other by Royalty as well as mileage.

The enactors in Williamsburg assured me that the buildings were original and some minor painting had been done to ensure they looked good. The streets were long and they led to the red brick state house at one end with enough room for horses and carriages. The church was located in the middle of the town again built in red brick. This is very similar to the Old Philadelphia structure of the streetscape. The homes are generally similar in both Williamsburg and Old Philadelphia. The colonial architecture is distinct with the two story houses and taverns and spacing between the homes on a walkway usually made of brick or cobblestone. The streets are laid out in a long crosshatch with one major boulevard down the middle of the town and various establishments located along the way.

Many of the public buildings such as government and colleges and large residences are Georgian in style very similar to Washington DC. The Georgian style received it's name from King George I, II, III, and IV of England. The style was very simple and many public buildings in America are done in this style. King George was from the House of Hanover in English Royalty[11], of German descent.

Slave quarters were nearby and the Black First Baptist Church was located centrally to the downtown in Williamsburg. One wonders what the slaves must have lived like in that world. They were expected to convert to Christianity and forget about family back in Africa tribes. This was very unrealistic and cruel without the economic power to visit their relatives. Curiously, there were no slave actors in current Williamsburg reflecting modern political correctness but historical inaccuracy. Most plantations are located outside of the city of Williamsburg. It seems the old south wants

---

[11] Georgian Architecture in World Book Encyclopedia, 1992, pp 135-136.

nothing to remain of the suffering of blacks under a system of slavery in similar fashion to South Africa's abolished apartheid system.

The Greeks originated the style of many of the buildings in Washington DC. The Romans then copied the Greeks. The British and French copied the Romans architecture during the period of enlightenment in the 1700's[12]. The American cities of Washington DC, Philadelphia, and Williamsburg were thus copied from the British and French. An example of the Baroque British[13] architecture in this style is St. Paul's Cathedral built by Sir Christopher Wren. He used Classical Greek and Roman architectural styles in the new creation in the 1600's. The large Rotunda and columns are central to the architectural style. Each column has a design on the footer and header that tell a story of the builder's priorities and the region. The US Capitol, Lincoln Memorial, Smithsonian Institutes, Jefferson Memorial, White House, and many other government buildings use this same technique with large wings built off the Rotunda and columns. Paris government buildings use very similar ideas from this period. The Arch de Triumph uses Baroque style Architecture. This makes perfect sense since Washington and Paris were designed by Pierre L'Enfant. Benjamin Banneker helped him with Washington DC.

---

[12] Herman, Arthur, How The Scots Invented the Modern World.
[13] Architecture in World Book Encyclopedia, 1992, pp 606-634.

7. The traditional CBD and Concentric Circles

   The traditional CBD (Central Business District) originated in cities in the

ancient world. The central business district was the area where businesses

marketed their goods in town. The market was inclusive of all the trade from

other countries in the city. The market sometimes included the temple. Jesus

was known for his speaking out against the bringing of money and bartering

into the temple of God. He turned the merchants tables over and said they

should not do this in the house of God. Today, the central business district is

the downtown area where the skyscrapers are located and major business is

conducted. In cities where there are no skyscrapers like Washington DC, the

government buildings are the central business district around the White

House and K Street. Manhattan has central park and Park Avenue with major

skyscrapers in the downtown area. Chicago has the businesses near the

loop in the downtown area that form the central business district. The point is

that it is a pattern of development and departure point from the middle of the

city in any city. Hoyt said that from this central point concentric circles around

the CBD radiated in an outward fashion to form rings around the city in which

certain development of suburbs and residential areas formed. The circles go

on until one reaches the outer band called "exurbia". Leap frog development

occurs along the urban highways. Urban sprawl also occurs as one traverses

the circles in Hoyt's concentric circle model from the inner circle (CBD) to the outer circle (Exurbia). This is denoted by most public planning as a problem and Harvard has a program called "Smart Growth" which helps the public prevent sprawl and leap frog development. Every urban planner worth his salt knows that by controlling local zoning laws and planning the new growth he controls the fabric of the neighborhoods of the city. The problem is that this strategy is often at odds with public opinion and private developers wishes and investments in neighborhoods. Profits should never take priority over preserving the environment and neighborhoods that already exist.

With urban rezoning and redevelopment comes the urban renewal that President Johnson hoped for in the 1960's when he funded drastic changes in most cities. The inner city needs reinvestment and this has happened in our time (1990's) with new money from young professionals returning to cities to build new homes for their families. Much of the research from the Brookings Institute is targeted at local Urban Planning. Local planning is also done at the county level in suburban counties around the country. This is the area where there is the highest growth in the nation. As families move up in income they have new opportunities to live in better homes. They invade neighborhoods and succeed previous groups of people who once lived in the older neighborhoods. We have seen entire neighborhoods of Italian, Irish, Polish, German, and other immigrants move up this way in American cities. Boston and New York are two examples of this on the east coast. On the west coast we include Japanese, Chinese, Vietnamese, and far east cultures as the invaders of the neighborhoods. People like to be around people like themselves, culturally. This holds true up to a certain income and education level. Diversity helps us to understand other cultures and accept all people as equals.

Hoyt's concentric circle theory has another more dastardly application. Say you looked at several cities collocated in the same region. The concentric circles from each city going outward from the CBD would eventually intersect like a Venn diagram at some point. Then there is the outer band called "exurbia" beyond suburbia where only the richest most affluent residents can afford the cost of living. Now, say that the cities in question are Baltimore and Washington and we are in the middle of a second cold war. The concentric circles map exactly with the kill radius of a thermo nuclear weapon or atom bomb explosion. Under this logic, the concentric circles become zones of protection the farther away one is from the inner city CBD or ground zero. In Washington ground zero is the White House and Capital Buildings and Pentagon. Some maps have concentric circles on them out away from the CBD measuring the distance. Once you are about 17 miles away from the epicenter of the would be blast (in exurbia), you are safe. Now consider that you have the Venn diagram situation of two cities nearby each other with overlapping exurbias. Take 17 miles from Baltimore's inner harbor and figure where that intersects with the circles from Washington DC, somewhere in Laurel, Maryland. The enemy foils the theory or dual intersecting exurbia protection by planning to drop a larger nuclear weapon in the middle of the two intersecting concentric exurbia circles and nails both cities in one try. In the case above, the logical mathematical point of reference between Washington and Baltimore is Ft. George G. Meade, Maryland, next to Laurel.

Another dastardly conclusion using urban planning in general is that the science and art of urban planning is now a perfect military discipline since we are fighting urban warfare. In planning the warfare and fighting, what more perfect person to tell what intelligence says about the enemy infrastructure than an urban planner? This is the same philosophy that the combat engineers used in all previous wars. They were the ones who knew the most about building and destroying bridges and other strategic and tactical targets. Sure urban warfare could be prevented with a neutron bomb that preserves the infrastructure and destroys the enemy but this is not official US warfare policy. Why, because the neutron bomb does not distinguish between friend or foe where civilians are concerned. It is pretty hard to determine this in any case in our current occupation of Iraq and Baghdad. Please see the chapter on Baghdad to see my assessment of that city with regards to specific details on civil engineering and other urban development projects we have undertaken since invading Iraq by US Major General Peter Chiarelli, a civil engineer and West Point Graduate. This has been very interesting to observe through the western media which almost covers nothing on it. So you see NSA and US Intelligence could really use an Urban Planner who knows Farsi,

Arabic, or whatever language they speak in Iraq or our other current enemies/terrorist nations.

8. Regional Urban Centers

John Naisbitt stated in his book <u>Megatrends</u> that America is thinking and
acting regionally. In America, we live in regions. The government has divided
it's agencies into regions. We commute to work in regions of the country.
The main regions are the Northeast Urban Region, the Great Lakes Urban
Region, the Mid Atlantic Urban Region, the Gulf Urban Region, the Texas
Urban Region, the Southern California Urban Region, and the Northwest
Region. All of these areas are major multi-metropolitan areas of urbanization.
People commute from cities in the region to other cities and congestion
ensues.

Groups of cities form a megalopolis. "Bo-Rich' is the unofficial name of
the eastern string of city and suburbs from Boston to Richmond. As the
suburbs fill in between major cities with sprawl and leap frog development the
local governments try to control zoning laws to ensure a mix of uniform
residential and business use in these areas. A new CBD is created in the
suburban areas where the filling in is happening. Suburban Virginia and
Suburban Maryland have taken on a life of their own. The councils of
government have too much on their plates to control all of the development
effectively in the megalopolis. As the megalopolis gets more urbanized over
time, farmland in this region is pushed to the west near Appalachia.

One political party pushes development for business purposes and the other party seeks to control growth and preserve farmlands and parks for future generations of children. One can see into the future of regional thinking and imagine how it will be in 2050 in America. We will have major urban regions and major rural regions with a population of more than 300 million needing to spread out. We are seeing the polarization in politics that reflects this already in the 2004 election. The southern or rural areas won the election over the northern and western urban regions. Urban dwellers actually have a different belief system that is more liberal and tolerant of different cultures than rural people have. Rural people think they are more moral because they live off the land and have less people problems such as crime and poverty. Cities will have to realize they have major voting advantages when they turn out to the polls en mass. A rich person in a city has to help his fellow man as he lives closely with him. The same is not true of a rural person. America was at one time a 95% rural nation. This has now been supplanted with a 95% urban and suburban society. The politicians of the future who will win will have to understand the urban and rural vote will make the difference. Politicians are already playing games with this by running diverse candidates who appeal to both the urban and rural diversity factors. I believe politicians sometimes change their views when they realize who they are addressing in public speeches.

As we become a nation of regions and the megalopolis, new technology will ensure travel between the major work centers. Satellite work centers and telecommuting will curb the growth in traffic patterns of congestion. Satellite cities or suburban residential cities are already developed that house workers who travel to other cities in the daytime. The government is sponsoring new work methods. The computer allows workers who understand them to tele-work effectively on a daily basis and this helps reduce traffic congestion. Some occupations are easier to do this than others, but many more will have to try it before we curb the pollution of vehicles on highways and lower the costs of maintaining buildings in the CBD. The key concern here is training older managers to accept new methods of work and electronic work products from home. Trust and verification is required for this to work effectively. Some managers will never be comfortable with telework.

9. Demographics and Population Studies

Demographics and population studies are critical to understanding dynamics and politics of major world cities. For instance, many capital cities have a large percentage of multi-national people represented in the population. This is because these people may have business interests in the capital city from foreign countries. Major working class cities may have neighborhoods of certain ethnic groups who do similar work. Industrial cities may have an influx of professional people and the fabric of the city may change over time. Many inner cities have experienced young urban professionals returning to the downtown area to beat long commutes to the suburbs. Politicians understand the population studies done by the US Census. These set congressional voting districts across the country and also tell us something about the population trends of the cities and rural areas. All types of data collected on the census forms is disseminated through our government agencies. Nothing that is privacy act covered is disseminated in these studies. Colleges keep census records for undergraduate student studies. The Maryland Planning Office uses census trends data to plan ahead forecasts for Maryland. This data can be quite useful in many applications in government service. The state and federal agencies work together to gather and analyze the data on cities, counties, and other rural areas. A good source of this demographic and populations data for foreign countries is the

CIA World Almanac[14], and Encyclopedias[15]. Major economic trends can be affected by demographics and population trends. This data is useful in determining marketing trends for companies in certain markets involving large populations. The data is free to the public researcher. In a democracy such as the United States where the majority rules, it pays to know the statistics about your people. Typical data collected is income, education level, sex, race, age, and vital statistics. Vital Statistics are those concerning life and death of the general population. Usually a single government office issues vital statistics annually. Many state police issue crime statistics and traffic statistics which are also useful to public agencies. Historical demographics cover many years trends in data. Projections are future predicted trends. This is the value of good historical data. Future trends can easily be generated using factors computed from past years and extrapolated into future years with a specific slope.

Spreadsheets are ideal for storing data cells of information both online and offline. The nice things is that charts can be produced that impress the customers of this data and really tell the demographic story of the population of that jurisdiction. Demographic data are not good at explaining individual behaviors outside the normal distribution curses and normal statistical methodologies. Population studies was once a required course in Urban Planning courses at University of Maryland before it was in the architecture

---

[14] CIA World Almanac Online
[15] World Book Encyclopedia and Encyclopedia Britannica

school. This program has evolved and is now under Architecture rather than

Social Studies which may make a big difference in the subject curriculum.

Figure 3. Population Growth of Top 15 Major US Cities

	1900			1960		2003	
Rank	City	Pop	City	Pop	City	Pop	
1	New York	3.437 M	New York	7.781 M	New York	8.085 M	
2	Chicago	1.698 M	Chicago	3.005 M	LA	3.819 M	
3	Philadelphia	1.293 M	LA	2.479 M	Chicago	2.869 M	
4	St Louis	575 K	Philadel	2.002 M	Houston	2.009 M	
5	Boston	560 K	Detroit	1.670 M	Philadel.	1.479 M	
6	Baltimore	508 K	Baltimore	939 K	Phoenix	1.388 M	
7	Cleveland	381 K	Houston	938 K	San Diego	1.266 M	
8	Buffalo	352 K	Cleveland	876 K	San Anton	1.214 M	
9	San Fran	342 K	Wash DC	763 K	Dallas	1.208 M	
10	Cincinnati	325 K	St Louis	750 K	Detroit	911 K	
11	Pittsburgh	321 K	Milwaukee	741 K	San Jose	898 K	
12	NO	287 K	San Fran	740 K	Indianapls	783 K	
13	Detroit	285 K	Boston	697 K	Jasksonvll	773 K	
14	Milwaukee	285 K	Dallas	679 K	San Fran.	751 K	
15	Wash DC	278 K	NO	627 K	Columbus	728 K	

Source: Time Almanac 2005, pg 205

Figure 3 shows the population changes in the top 15 major cities in the United States since 1900. The emergence of Los Angeles since 1900 is evident in the 1960 and 2003 data. Many of the cities listed in 1900 have lost total population since that time and have dropped off the list due to migration to other parts of the country and suburbs. The southern cities have recent gains in population in 2003 as people moved to the sun belt. The life expectancy is now 74.5 years old at birth in the United States. One can surmise that many citizens are moving to the sun belt to prolong their life in fair weather as well as lower the economic cost of living.

It is interesting to note that every city in the 1900 list shown has a major league baseball, basketball, or football team in 2003. The owners knew where the money for tickets for each franchise would come from and surely had this population data shortly after the civil war when baseball started in major cities. The Congressmen of those districts must have fought for teams in their cities to keep their constituents happy while they ran the country and offered tax breaks and incentives to the major league owners. Baseball in the cities was a way of life to entertain people and control crime.

Let's take a look at the 2003 list of cities in figure 3. New York is a far larger metropolitan area than the population that is shown and the surrounding counties have a far larger population than is listed here. This is true of many of the largest cities in America. One of the questions is when can we tell how a city infrastructure has reached critical mass and maximum saturation and population density? The growth factors for the 20th century suggest that in 2050 we may have 1.5 as much population in each of these cities.

10. Socio Economic Status

Socio economic status is an indicator of wealth in the private individual's life that is tracked by the government census forms. Adam Smith would have been a proponent of socioeconomic status as a means of nation building and national wealth building. In other words, there are "haves" and "have nots" in society and socio economic status is a differentiating factor in many other aspects of American life. High socio economic status may enable a person to participate in a certain job market or attend certain prestigious private schools. It is possible for someone to change their own status in America borrowing from the American Dream to improve oneself and one's family. Generally, we are raised in certain values as children that determine our status. College bound children who complete their education will be rewarded with increased salaries over the long term. One day a college education will be the standard for all Americans rather than a selected few. President Clinton understood this dynamic and was very effective at enabling the youth of college bound America by making it affordable to lower income students with good academic records through the Hope scholarship. By improving our National education levels we increase the socio economic status of mankind and modern America. The Internet in every house was another major effort that would serve to improve education levels of our population and cross the "digital divide" between a nation of "haves" and "have nots". The American Dream is still achievable by young people and they need to know this rather than being scared about the lack of social security funds. Building economic security is based on thinking a certain way that is highly available to all of society. There is no color nor ethnic barrier in high socio economic status since high education is one of the factors. It is true that higher education in America is a generally European Institution and may be racially biased. However, many schools such as University of Michigan are leading the way in recruiting minority students in the future to ensure equal representation and opportunities to access the education system. The various cultures of America must make sure that they tell the children to keep the faith in the education system. This is the crisis in the black community that Bill Cosby speaks about. He has been a proponent of excellence by all people of all colors in the academic system of America. It is clear some people have distinct advantages by the way they were raised and the things they believe to be true in the educational system. Look at any successful people in politics and government and they will tell you that good schools got them where they are today and hard work. They also met good people who helped them along their journey to high socio economic status.

11. Education

Education is the great equalizer. It helps bring people out of poverty and builds a better nation of wealth as Adam Smith envisioned. It is the cornerstone of advanced societies and goes back beyond the early bible when man wrote on walls in caves and first started communicating with grunts to each other. Good public education was found in the Roman Empire and Jewish nations. The temple in the Jewish faith is a powerful reminder of learning for young boys who read the Torah. Muslim people have strong religious beliefs and it is a myth that they have few historically great scientists. America has a great education system that many around the world covet because we also have opportunities to match the education system. Some countries have totally government sponsored education systems like Great Britain and Japan. These are very affordable systems for these countries and we have much to learn from them both in terms of student achievements, funding, and student dedications. The best and brightest are earmarked for further state sponsored education. In America, the rich families send their children to rich schools as much to make future contacts as to learn new information. Christian education in America leads us to many good conclusions about ourselves and our openness to others. Christian mis-education prevents us from loving our neighbors and our enemies as ourselves. Parents need to educate their children in Christian as well as secular ways of the world. Whole branches of science have branched off of Christian education because people did not see where they converge rather than diverge. Darwin's Theory of Evolution is an example of this. Darwin hated the Christian education system and he boarded the ship "Beagle" to find freedom from it. Isaac Newton, on the other hand, used a Methodist Church for many of his experiments before he wrote his book "Principia". Science can prove biblical prophecy and some prophecy can be very scientific. This may be seen in the events of Moses life. Scientists believe a volcano erupted about the time Moses parted the Red Sea causing the waters to draw back. They also believe Moses may have spoken to aliens on Mount Sinai and received help from these aliens. True faith does not require a scientific explanation, but modern man needs science.

The best educated people will be the most flexible and understanding of others as well as themselves. This we call religious tolerance. It is a timeless and cross religion principal of freedom in America and all modern democracies. Public educations are county and city run systems so we must support our children in their need to discover how they fit into the systems. Moving to a good school district only solves a part of the equation. Engaging our children is thinking at home and in fun activities with parents solves the other parts. In my mind, this is coupled with good marriages that last a lifetime between caring parents. This is the best building block of the family that supports the education system.

It was calculated that a major university[16] produces $6 into the economy for every dollar invested in it and that this is the highest ROI of any institution in America including business and military. SAT Scores are going up over the years since 1960. But, so is the tuition cost of a good college education. US News and World Report creates a list of top colleges every year and ranks them for value and puts them in three tiers of ranking for excellence. Tier 1 schools require the best SAT scores and so forth. Today, we have DETC online colleges and everyone can go to school in non-traditional ways that were never available before electronics communications. This was a gift from president Bill Clinton who was a great education and electronic digital human being who wanted to knock down the digital divide in America between rich and poor people. He truly had insights into improving education for all with the Hope Scholarship and increases in people attending community colleges. There were also new ways to acquire funding under president Clinton and teachers felt that he was on their side as he had been a teacher during his career. The current Bush administration is strongly lacking support of teachers and teacher unions. One has tried to make it on academic integrity and the other has tried to make it family name, friends, and a higher moral calling. The difference really shows in the education policy differences between the two presidents. One blames the quality of teachers (Bush, Harvard Business School) by instituting draconian certification requirements and one improved the overall educational environment (Clinton, Rhodes

---

[16] University of Maryland, College Park, President CD Mote webpage, 2005.

Scholar, Oxford University) at home and in public. This really shows where these men have been in their own academic careers and what their public opinions and policies are towards others.

Figure 4 shows some enrollment data about colleges and universities in the United States. The total enrollment is 15.9/284 or 5.28% of the total population of America going to college in 2002-03.

Figure 4. 2002-03 US College and University Enrollment

Institution	Number	Enrollment
Public 4 Year	631	6,236,455
Private 4 Year	1,835	3,440,953
Public 2 Year	1,081	5,996,701
Private 2 year	621	253,878
Total	**4,168**	**15,927,987**
Undergraduate		13,715,610
Graduate		1,903,730
Professional		308,647

Source: Time Almanac 2005 p.321

Figure 5 shows the number and percentages of students awarded degrees in America. Almost 3% of the total population is receiving some type of degree. America needs to do better than this. We are also recruiting top college prospects from other countries as we always have to improve our academic ranks.

Figure 5. 2002-03 Degrees awarded

Degree	Number
Associate	595,133
Bachelor's	1,291,900
Master's	482,118
Doctorate	44,160
Professional	80,698
Women	56.3%
Full-time	59.3%
Minority	28.8%
Foreign	3.5%

Source: Time Almanac 2005 p.321

12. Satellite Cities and New Towns

The list of satellite cities and new towns is growing fast since World War II. Greenbelt Maryland was a government project in the 1940's and is one of the first new towns. Today, residential neighborhoods are locating in rural counties to create satellite cities sometimes called "Bedroom Neighborhoods" because they only have enough services and businesses to support commuters who work in nearby cities and travel there daily by public and private transportation. Reston, Virginia and Columbia, Maryland[17] are two new towns of this caliber near Washington and Baltimore metropolitan areas. Both cities have grown up a lot since they were first chartered and now support a myriad of businesses in their own CBD and downtown areas. The concept is new in that shopping malls are at the center of town as the meeting place. Churches, businesses, and other meeting places are also located in the CBD. Reston has an international building with many businesses. Both towns attract people with amenities that go beyond the normal towns such as golf courses, community centers, athletic complexes, and swimming pools. I often find the design of these cities similar to the communities that are formed on military bases in America. These are small survivable communities that have everything they need provided on the base for those families in the service who live there. This is a very protected community with it's own police force, own firehouses, own schools, and gymnasiums. The similarities are not by chance. The contractors who build these neighborhoods use similar

---

[17] The Columbia Flyer and Howard County Times newspapers.

concepts for both. In the case of the military community, all the people work for the same boss, Uncle Sam. In the new towns, people work for a variety of employers who do not provide housing or services.

The satellite city of Londonderry in Ireland was one of the first cities that was located outside the main city and commuters traveled to and from the city by day. This town was revolutionary in that it was a first attempt at this type of sub urbanization in the western world. There were only services essential to supporting the population located there, no CBD. It was a bedroom community. The videos were shown in urban planning courses at Maryland. Many civil engineers know about this city. The American Society of Civil Engineers has dome extensive work in Great Britain and plaques are located next to the projects stating they were the ones who built these many projects. The average distance to a major city from the satellite cities is around 30-45 minutes on a super highway at 60 mph. Columbia, Maryland is within this range of both Baltimore and Washington. Reston, Virginia is within this distance from downtown Washington DC. Both of these were financed by private companies and sold to private citizens and investors. The return on investment has been significant for those who took the risk of moving there as property values have soared in these new towns due to high demand by all segments of society for a better quality of life.

13. Villages and Neighborhoods Concept

The village movement is one that takes us back to the 1700 century
mainly spearheaded by developers. Columbia, Maryland is a place where
villages and neighborhoods have taken hold. Each village center has a
shopping mall. The main shopping mall is at the town center next to the CDB.
Each village is different yet the same. The concept provides for a
homeowners association for all of Columbia partitioned by the villages. This
management service is provided at cost to the homeowners and is the price
agreed upon when the home buyer is at settlement. Each village center has
it's own identity in the local fabric of the community. Each community has a
school, swimming pool, churches, shopping, and other amenities. The
developer of this concept was James Rouse who helped design and build
Columbia. It is the closest thing I have seen in America to the military base
concept of enclosed communities with everything they need. The housing is
high end and mixed use or at least it was intended to be as such by the
Rouse vision. Rouse recently sold it's interests in Columbia to a company
from Chicago.

People congregate in this form of neighborhood with other friends nearby
in the same and other villages. Interesting car pools form and people network
from all over the various villages each taking an older view of dealing with
people. At the lake in town center there is the people tree which represents

the diverse people of the city whom live there in the small manageable unit called the village. It works quit fine until it comes to regulating the infrastructures. The problem is that owners are asked to police other owners whom may have reasons for avoiding covenants that apply to everyone. A form of communism ensues where one has no rights to alter ones own property and one must keep up with all the lawn work and gardening or else neighbors will report the ones who leave the job until a later date – all to maintain property values. It is a good system in design, yet practically there are many complaints and bad things about this type of semi-communal living. Sure properties are kept up by the able bodied even for those whom can not do the tasks required. But it is frustrating at first to the newcomer whom wants to modify his house the way he wants. He must obtain permission from his immediate neighbors by the law of the neighborhood that he signed when he purchased there. He can not have a boat or extra vehicle in the shared parking of townhouse units. He can not do any improvements without having signoffs by neighbors, except for repair work that is similar to the existing structures to maintain status. As one woman said she felt like she was in Nazi Germany when her lawn was 6 inches high. She was old and could not mow the grass herself. You get all this regulation and you pay an extra fee which amounts to another tax. I resigned my commission as architecture board chief later that month and moved from the neighborhood to retain my own property rights in my new development where I did not join the homeowners association for fear they might charge extra. My village experience was not lacking in high quality living though and I enjoyed the large number of amenities in each village that were close. The people were also very nice for the most part created by the fact that there were walkways and parks for the kids (tot lots) throughout the development. The developments were really well thought out for family living. This was greatly appreciated by everyone who lived there.

14. Parks and Planning

Parks and planning is a function that creates a nice, friendly community where people are welcomed and can rest and find peace and relaxation. We call this "Context Sensitive Design" in today's engineering world or a concern for the way we engineer all aspects of our communities. Parks and planning were a concern of Robert Moses who had many designed in New York City. This idea was then taken to other cities in America. The United Kingdom uses parks very efficiently in urban/suburban/rural city designs. They are a place of comfort and provide a nice place to take family and friends. Central Park, Rock Creek Park, Patterson Park, Druid Hill Park, are all places one can go to enjoy the outdoors within the city preferably during the daytime. They also run golf courses which are another designed human friendly place in cities. Parks and Planning in the National Capital Area is done by MNCPP. They coordinate all efforts at building, transportation, and running facilities in the parks. Where would we be without nice parks systems in the various states?

The western states have some really great parks like Grand Tetons, Mount Rushmore, Grand Canyon, Yosemite and others. These parks are managed by the US Department of the Interior who regulates them. Recently, they have been under assault for teaching God and creationalism rather than strict science. Both have merit. The local parks and planning commissions uses the master plan and provides input to the local planning and zoning people on the plan when it is developed. It can be contracted out to be revised. It is often called the Comprehensive plan and can be done for counties, cities, or local municipalities. The community gets a chance to comment on the plan and process under testimony to local councils. This plan is rewritten and updated every certain number of years in most jurisdictions. It is the map of what is trying to be accomplished by local government planning officials. They are very detailed and allow for expansion. The plan is accompanied by the budget which is also used by parks and planning and every other agency involved in community building,

Some of these include highway agencies, energy utilities, firemen and police departments, schools and education, and churches infrastructure. Concentrated planning by all of these builds the fabric of the cityscape infrastructure, and yes, there is urban redevelopment too. Parks are now developed in new towns where they are the needle sowing communities and diverse people together. An example is Columbia, Maryland. Parks and planning works closely with the county to ensure that the city is context sensitive and ergonomically nice for it's citizens. The city is not incorporated as such but remains a part of the county planning process. This is unique for such as a large city. The government structure is called a "Charter" form of government run by a county executive and council of 5 district councilmen all over the county including the unincorporated city of Columbia. The advantage of this is that the county can make plans using all the developers for the whole county and maintain a high quality of life across all tax payers. Logically, the people are mixed income and those paying the developers more will pay more taxes into the revenue base of the county. An association fee pays for the amenities in parks and nice areas designed for the population by the developers with newer building techniques that are people friendly.

15. Transportation

   Transportation is a key element of every federal, state, and local master

plan developed concerning city management. Without transportation there is

no business or economic development in communities. The very urban

sprawl that many complain about is the natural economic business that

develops when major arteries are built by highway departments. Smart

Growth counters this sprawl. Public transportation needs to be provided for

urban and suburban areas linking the CDB to the satellite cities. The workers

need to have multiple ways to get to work. Today, telework is keeping down

the traffic densities by helping people work from home. Only those who really

live the farthest or have health problems are first candidates in most agencies

and work places. The idea will catch on better and there will be a day when

many people stay at home to work at least half the time. The problem with

telework is that one misses a lot by not coming to the office every day in

terms of human contact and relationships. Computers can be setup to

transmit work to the workplace very easily and congress supports the idea of

national telework. Maybe some day they will have a national telework day.

This program also helps reduce pollution as well as traffic flows and

congestion.

   Corridors of traffic provide passers by whom stop to do business at local

establishments along the highway. The local zoning boards control which businesses get local access driveways directly to the highway and they also control the infrastructure such as sidewalks, parking lots, shopping centers, trees in the medians, median strips, traffic lights, pedestrian lights, and almost every other facet of the corridor. Private businesses pay taxes to have the infrastructure maintained by the government agencies managing the highways and local roads. This is usually done at the county and state levels.

Transportation is a federal agency with a presidential cabinet member. But it is also delegated to the states and local governments to manage on a daily basis. The US DOT secretary is currently The Honorable Mr. Mineta. He coordinates his administration with the FHWA, FMCSA, and NHTSA agencies to ensure a successful transportation system. The National Academy of Sciences Transportation Research Board is also an advisor to the US DOT. There are many good people working hard at these agencies to ensure the mission of US transportation is met during stressful times of terrorism using our national transportation system as a weapon against us as in the world trade center disaster of September 11, 2001.

16. Financing Urban Development

   Urban redevelopment and urban development take financial resources
that are not always immediately available to governments. This money often
must come from the residents and commercial businesses ultimately.
Sources of Funding includes property taxes, building bonds, municipal
bonds, federal housing monies, section 8 housing grants, local and state
income taxes, gasoline taxes, and other fees.

Municipal bonds are offered for projects that are building projects in cities or
jurisdictions. They are paid back after the project is initially funded.

Federal funds can be specified from legislation by specific appropriations bills
for certain purposes of building cities and urban projects for people. State
funding can also be allocated to priority funded areas as called in Maryland.
Local funding can be used for county level projects and specific projects in
certain locations. These building funds come from a variety of tax sources.

HUD funding is funding directly from the Department of Housing and Urban
developments such as low rate loans to first time home owners, HUD loans,
and VA loans with reduced interest rates and government guarantees. The
HUD 1 form is the form used for this type of loan.

Private developers can fund certain projects through various loan mechanisms from people buying into development projects at higher economies of scale. They will usually use federal monies and programs as well as conventional financing programs for people interesting in moving into residential neighborhoods. Private developers can also help fund commercial properties through business loans for commercial properties and the federal government business loan programs.

Commercial taxes help defer costs associated with commercial zoned areas. The government uses the money to apply to maintaining the properties in these areas or zones. This would be a local government function. Sales taxes and tourist taxes are two types of commercial taxes that can applied to help build commercial areas.

Property taxes can be levied on commercial or private residences. They will be used for financing the services for the properties. They are assessed annually and usually are paid through normal residential and commercial mortgages process.

Gasoline taxes are levied at the gas pumps in the various states. They find their way into the revenue system of the government who levied the taxes, usually states.

Fees and Tolls include special fees such as vehicle registration fees and tolls are used to pay for transportation related activities in the urban and rural environment. These are also levied by states laws.

17. Urban Redevelopment

In the 1960's, President Johnson created the Great Society. Part of his vision was the urban redevelopment of downtown and residential areas of America's cities. He funded projects that were rebuilding the downtown housing areas. He provided hope to inner city people but actually displaced quite a few also. Affordable housing was always a goal of the urban redevelopment projects of this time period. The actual goal was not always met. Some of the shining stars of this time period in history are the new towns of Reston, Virginia and Columbia, Maryland. This was truly a time when people became more open and understanding of other people and the civil rights era gave rise to the economic and other struggles of minorities in America.

The primary reinvestment was in downtown areas. Urban renewal became a bad word in many cities. But when the project finally completed there was always greater hope for the future for a mass of people who needed a hand up. The primary investors where people who felt their neighborhoods were important enough to save through investing money in development through banks and other federal funding. During this time neighborhoods were experiencing flight to the suburbs by residents who no longer wanted to live in the downtown housing areas.

The renewed commitment to civil rights by Democrats changed the way we do business in America and made our country stronger. Housing laws protected all people and made it illegal to discriminate in housing. Yet, in many parts of the country the changes were slow to be instituted. Interracial marriage became legal in many states in 1968. Old segregated institutions started to crumble under their own wickedness and exclusionary tactics. Schools and neighborhoods began becoming integrated for the good of all concerned.

Higher education in cities became the way to improve the quality of life for all Americans across the board. Veterans from Vietnam were returning to schools to use their GI Bill. Children of veterans also went to schools. The government advertised college education as a reason to enter the military and people did.

Downtown churches banded together to continue making a difference in the lives of their citizens. The faithful returned to their churches to help building projects for the poor. Where government could not make changes fast enough, the church establishment did. Rev Dr. Martin Luther King led the way in helping people of all colors realize they did not have to settle for less than a high quality life in modern day America. He opened many people's eyes of all colors, races, and nationalities. Some shunned him and did not want change. But the force of faith and God was too irresistible for America. As the war in Vietnam raged, we tried to solve these problems at home. Many became disillusioned and many more protested government policies that were not helping the people.

Inner city revitalization was accomplished in many cities. By studying the great Urban Planners such as Robert Moses, one can get a feeling for how various people affected the life in some of America's great cities. Robert Moses made changes in New York for 44 years that reverberated throughout all major cities in America. He tried design changes that led the way in other cities such as downtown parks, public beaches, interstates through city neighborhoods, and large building and construction projects in formerly unwanted lands.

Increased funding for public improvements to cities helped fuel the investments in the landscape of the city. People bought municipal bonds and helped fund projects. Public perception was one of great social and cultural changes taking place in America. Reality was that engineering and urban development was changing the very face of the city where the people lived.

Civil engineering projects improvements made an impact in inner cities. Larger buildings were being built such as the World Trade Centers in 1972 due to new construction techniques. Skyscrapers and bridge designs became bolder and bolder pushing the limits of beauty as well as design within the operating budgets of major cities. Commercial and private buildings began taking on radical new designs. We saw a solar energy house develop and in the 1970's the energy crisis caused many designers to come out with new ways of fueling homes and vehicles that were never before tried. The space race to the moon affected our expectations of new designs also. We were now in the Age of Aquarius and new aerodynamic fantastic designs were a thing of the present.

The Brookings Institute has always been a think tank on many of the public policy issues of this time frame and currently. They have many holding and readings on the topic of urban redevelopment and the history of this time period. It would be wise to browse the holdings of the Institute for any ideas on future urban redevelopment. The idea that young people are willing to move back downtown is a rather new idea that merits attention. Young urban professional who can afford better housing and actually love living in the city are being attracted back. Also many government agencies have started programs to pay a bonus to employees who live in the city or jurisdiction where they work. This has the net affect of improving the tax base and economic situation of these areas of urban redevelopment.

18. Suburban Life

The population growth trends in the 60's and 70's were to move out of cities into the suburbs and this trend still goes on today as people can afford new housing at higher costs. There even seems to be a trend with an aging population of people movement to the south or sun belt to both live on less income better and preserve health in warmer climates.

Suburban invasion and succession has always occurred in these parts of cities. Entire neighborhoods were created of the same type of people who immigrated to America in the large cities. Like minded people lived close to each other. Today, with upward mobility, families of all backgrounds are living in various subdivisions around the suburbs of America. New immigrants still congregate together but after several generations they will disperse once they have economic power. The immigration issue in America has always been one of concern and as we grow to almost 300 million in population we are becoming a strained society on our financial resources and abilities to take care of all citizens. True there is a lot of land left undeveloped in the western United States, but some of this land is uninhabitable.

There is a higher perceived quality of life in suburbs which is why people wish to move there. This can be measured by decreased crime rate, safer neighborhoods, better schools, and more peaceful existence. This will surely make the suburbs attractive to people for years to come.

At what cost is this perceived quality? The taxes are higher in suburban areas and families are more integrated into the fabric of the neighborhoods and human systems along the axis of work relationships.

Mixed housing for all exists in modern suburbia and many new towns and new cities are trying to ensure all incomes of people are able to afford the lifestyle of suburban living.

Stamping out poverty levels in new suburbs is the best way to defeat poverty on a permanent basis here in the United States. When people are offered a better way to live and shown how to accomplish this and given hope, they will excel to the next economic level.

New towns have become our hope as the new suburbs of the megalopolis. New town's are constantly growing and becoming more diverse and competing interests are legislatively acting in their own economic benefits. The growth in new towns is such that they are redefining where they will house more people and at what expense. Land is a prime consideration and is at a premium in new towns. An example of this is Columbia, Maryland who is experiencing growing pains as a 35 year old new town. A new development corporation has taken the lead on developing old properties that were used for other purposes in order to provide more housing and make profits. There is currently controversy over weather to develop the downtown area open areas near a local amphitheatre or leave it alone.

Where are the best suburbs in America? The answer to this question needs more study. Surely there are young families who would benefit from the answer. Would they be where the most jobs are or where the best living conditions are or both? The ideal situation is one where the family can grow and have economic opportunities as well as social and housing opportunities they can afford.

19. Regional and Local Comprehensive Planning

Planning is the most critical management function of all of the seven classical functions of management. Without planning there can be no implementation of plans. The master plan or comprehensive plan is one that encompasses many elements that create a higher quality of life for residents of the local jurisdiction they live in. The statewide master plan discusses the statewide initiatives that the governor and other political forces are trying to shepherd through the legislation into law or just into reality. The typical plan can be anywhere from 50-150 pages in length and is very inclusive of the type of changes in infrastructure that need to take place. Developers depend on the master plans to ensure they are seeking the right business opportunities to build communities and commercial enterprises. The local governments use planning as a tool to manage growth in the future and ensure every ten years there are changes made to reflect the real changes in the environment. In the United States there are more than 3000 counties and each of them needs a master plan. The master plan is a public document and can be used by all parties in managing the growth of the county/state. The document outlasts any one politician or his term in office. The smart urban planner will have a group of master plans for an entire region so that he can coordinate the future builds of various corridors and areas. He will try to minimize urban sprawl and leapfrog development in semi urban areas where new growth is a candidate for success.

Changes to the master plan are legislated or commented on by the public so that changes can be made that are logical. The details of the master plan are supported by the budget documents that pay for the projects outlined in the master plan process. The writer of the master plan needs to have a broad overview and understanding of all the elements of urban development local to that jurisdiction. In today's world, the magic of telecommunications allows us to download online internet versions of master plans and store them for future reference usage in transportation planning, commercial planning, residential neighborhood planning, local taxation, and redevelopment spending. Any citizen can purchase a master plan and see how his government plans to develop where he lives in the future. This is a worthwhile venture for the first time novice. The cost is about $20 for the printing fees for the Master Plan[18] in most areas. It is a wise investment for those in real estate and residential home sales who are project managers. It is also wise to acquire the budget that accompanies the master plan and pays for the improvements.

---

[18] Howard County Government, Zoning Division.

20. Rural Development and Protection

Rural development and protection is important because there is need for development that is structured and farmlands are vanishing. Once in America 100 years ago, farmlands were 95% of the total economy and urban service areas were only 5%. Today, this has shifted and only 5% of the economy is focused on agriculture while 95% is focused on service areas. Many state land grant colleges and universities started as agricultural colleges. Harvard's program for preserving and developing rural areas and suburban areas is called "Smart Growth'.

Sub developments and new communities use smart growth in building where farmlands once existed. These new communities are built to conserve land and use all the available resources to make them "context sensitive" in design. Developments must have adequate transportation for the population, adequate sewer and drainage, and adequate shopping areas and support services nearby. The commute to work then becomes a question of longer time in the car and higher quality of life tradeoff living outside the city.

Priority funded areas are areas where federal and state money is spent on major projects to bring urbanization to the areas needing development. Projects are funded in these areas through legislation at the local level. Projects may be developments or new highways. Both bring new economic benefits to the community.

Lack of profits in farmland development are partly the cause of farmers developing their lands in the exurbia areas of cities. When farmers can not make a profit on crops or no longer have a labor pool they are forced to sell their property.

Rural Highways development is undertaken by state departments to connect various urban and rural areas together and relief congestion as well as increase the local economy by shortening travel time and offering new businesses a place to co-locate on the highway. Rural areas are then developed and residences can be built or commercial areas to create a suburban feel to the environment. Environmental impacts statements are also done to ensure a congruence between new highways and the environment in the long term.

Access to major urban areas is the primary reason to develop the new rural areas and give high quality places of residences in a growing market of housing. The housing market is still a huge business sector that has positive returns for homeowners and is growing in America. It is estimated that 72% of Americans own their homes and the president wants this number to go even higher to help stabilize the economy.

These old rural areas become the new exurbia and new suburbs to major cities as development is underway outside city limits and between major cities. Leap frog development was the term used when random places were built. Urban Sprawl was another term to describe random development. By planning the process at the local level in a comprehensive plan, the development is controlled and becomes part of Smart Growth.

Hoyt's concentric circles are moving outward from the city and overlap between major cities in close proximity. This is what causes the megalopolis. They are also called SMSA's for statistical purposes of the Census Bureau.

Interconnected megalopolises then go onto then next stage of becoming a large regional entity of mixed development. We now see rural America and urban America connected by these largest cities, suburbs, and exurbias. Like a Venn diagram an exurbia may be in two or three megalopolises.

Providing small town environment in America is important to help people grow families at a rate that does not incur harm to the children and encourages friends and neighbors to become closer partners in the exurban and suburban experiences in modern day America.

This approach is towards a balance of economic and moral values. If one believes that rural America has better morals than inner cities then Smart

Growth is a perfect tool to grow more rural areas into exurbias of economic wealth as well as moral clarity. Education resources are required for these areas to ensure both of these goals.

Protecting rural preserves is the final goal of developing rural areas and this country has a strong National and State park system designed to protect historic conservations. In fact, in developing New York City, parks were used to give a higher quality of life to the residents.

21. Urban Geography

At the economic heart of it all is the CBD or <u>Central Business District</u> of
the city. Most cities are built around the downtown area known as the CBD.
The CBD can be built around a harbor or near the water as it is in Baltimore
and Chicago. It can be built on an island like Manhattan and Honolulu. The
sure thing is that the CBD is where goods, services, and money exchange
hands. Also government works in the CBD and often locates there for
convenience.

The <u>population centers</u> in the city are located in the outer circles of the
residential areas near the CBD but also near parks and suburban rural areas
where children can be raised and there is plenty of open space. There are
some residential areas in the downtown area but these are very expensive.

The <u>housing centers</u> form several concentric circles around the downtown
CBD. New housing areas form on highways and routes where access is easy
for the public and residents. Residential zoning adds to the tax base of the
city for property taxes.

<u>Commercial zoning</u> also adds to the tax base of the city through property
taxes and business taxes. Commercial zones and residential zones are

ideally separated by buffer zones. Sometimes residential and commercial zones collide together.

The education centers are sometimes located downtown and others located in the suburban areas. They always need parking. The education centers are crucial to ensuring that the business sectors have the right amount of trained staff. Many people will make trips from the business districts to the education centers after work hours and during the day.

Parks and recreation are important because they are the buffer zones that keep the various other areas in the city separate. City golf courses are nice because they also serve a function for the residents as well as provide nice parks for all. Parks where people can congregate in the summer are nice because they remind us that we need time with nature and are not separate from nature.

Highways and freeways are the access to and from the external parts of the CBD and city. The super highways of today provide high speed access to downtown areas and make the city closer to the residents who live in the outer suburbs. Working downtown is easy in this new highway environment except for the traffic jams and congestion experienced every day.

Subways are the mass people movers that most cities rely on to get people into the downtown area from the suburbs for work and day activities. Many cities have fine subways that are models of what we would like to see in all cities. Inexpensive rail travel is a blessing to the population. They do not need to pay parking fees or find parking garages while at work. The Washington DC metro is a recent subway system that is an excellent example of a modern subway with very nice designs. The Paris subway is an example of one that has been around for a hundred years and is in need of repair and redesign.

Airports planning is done with the idea that easy access to airports can be done in the local suburbs of the cities of today. The downtown airport is a thing of the past. The reason for this is the need for space for landing the larger aircraft and super jumbo aircraft on longer runways. The downtown runways are limited by the amount of space they can expand due to restrictions around the airport. FAA noise restrictions also prevent more airports from being located downtown.

Most military bases are located outside the city limits of the nearest city. The rare case is where a base in located downtown such as Lowery AFB and Bolling AFB. In older times the military base might have been the town, but now the bases are located where they have space resources to conduct military exercises. The federal government ensures they have enough space to conduct business without interference from states or local governments. The people living on these bases are also part of local traffic patterns and shopping patterns in the suburban landscape and planners need to consider them.

For the most part, urban geography can be predicted if one looks at the patterns of development in the cities around the world and in America. The component parts mentioned above are the parts of the city that make up the urban geography.

22. Urban Politics

Urban politics is always an element of the city. Big bosses run the city and the city manager is usually the one tasked with urban development and building project management. An excellent example is Robert Moses, the city designer of New York City for 44 years. He was a republican who mastered the ideas and concepts of governing projects that benefited the masses. He had vision as a politician. He also ran for governor, although he was unsuccessful. He worked for many city mayors including La Guardia. He was always aware of the money and legal actions in congress and he was the first to use this knowledge to his political advantage to build worthy projects.   The elements of urban politics include understanding the local people of all the neighborhoods and listening to their concerns before major projects are undertaken. The book Urban Villagers was assigned in Urban Planning courses at University of Maryland in 1978. This book described the lives and dynamics of the Italian West Enders in Boston, Massachusetts. Another book on the official reading urban politics reading list was Patrick Moynihan's The Melting Pot about life in New York City. New York politics lends itself open to criticism because New York ranks number one in population in America. Lately, ex-mayor Rudy Guiliani wrote a book on Leadership in which he discusses his principles or lack of principles in governing New York. I put the book down when I read the section on bribing someone who stays bribed. I have never thought of fair pay for fair work that way at all. President Lyndon Johnson had many urban policies during the 1960's Great Society that were designed to create a better urban environment in the United States. These are all chronicled at the Brookings Institute think tank in Washington DC. He basically poured millions of dollars into inner city urban redevelopment in hopes of redeveloping the poorest neighborhoods of American cities and improving quality of life. Unfortunately, the Johnson tapes on CSPAN also aired a Lyndon Johnson who was a racist Dixiecrat (by his own misuse of language on presidential tapes) who signed civil rights bills to make himself look good, not because he believed in them. But he did sign them and set forth the civil rights and human rights we all enjoy today. I guess if you do the right thing for the wrong reasons that counts for something. Richard Nixon cut back on money earmarked for cities in the late 1960's and early 1970's. Today, politics includes all fabrics of society and rightfully so. The protest of the 1960's showed teenagers they did have power to influence the government from outside and on college campuses.

Radical groups like the KKK and Black Panthers had definite impact on public urban policy with regard to civil rights and equal rights for all people who make up the urban environment and those minorities who need a hand up not a hand out. America became much more urban by 1960 than it was in 1900 by population and it was actually a good thing society became integrated on a whole because we were living in closer proximity than the colonies and new land was not to found. We need civil rights in order to respect each other in close quarters in future generations. Mixed density urban housing was not increasing fast enough for market demand. Affordable housing is still an issue today in most urbanized areas.

These same issues are with us today in local politics as politicians deal with the urban and rural counties and big cities from whom citizens are fleeing. Washington DC was long known as "Chocolate City" and it's "Vanilla" suburbs in regards the phenomena called "white flight" from the inner city. Today section 8 families move out of the inner city of Baltimore to areas in local counties where they can receive subsidies for affordable housing. This trend has reversed itself in recent years in some cities with young professionals and civil servants moving back into the cities with certain incentives for living in the district where they work. Property rights are always an issue in local politics as are corporate rights in the city CBD. In some places like Bethesda, Maryland the buildings have air rights that extend straight up in the air where they are built. This is similar to the mineral rights that a rural property owner enjoys. Bedroom communities usually have homeowners associations that can be very powerful and influential and sometimes act like another taxing authority. The development corporations are the ones who have control over the building at the local level with the politicians and zoning boards who zone for building and let contracts for competitive bidding for development projects. Corruption can be easily found in this area of government operations as many have been tempted to pay family members as contractors or for themselves once they leave government service. In one instance, a man who was a contractor became a government representative as he was unpleased with current policy makers. This was totally legal with certain conditions of service to ensure integrity in the process.

Urban politics is also an element of foreign cities. On a recent trip to Okinawa, Japan it was interesting to find that in the capital city of Naha, there were at least 6 political parties including a small communist party. The largest party was the Liberal Democrat part. This may be because they are such a new democracy after world war II. This reminded me of Great Britain and London where they have three major parties - Liberal Democrats, Labour, and Conservative (Tory) parties. It is not practical to ask anyone what party they are form, but can be gathered from the newspapers and the company who owns the local newspapers. They usually have a political slant as their publishing philosophy following the owners desires and direction. I have found many newspapers in Scotland to be very much more liberal than here in the United States and they do not make a big deal out of sex and act more moral than us. This is also reflected in the cultural art in the Louvre where nude statues are commonplace and not frowned upon like in the United States but rather appreciated for human art. America has some growing up to do in this area.

23. The Coastal Megalopolis's

    When mapping most of the world's population a large percentage is very close to shorelines in major cities. This is not by chance. Man needs water to survive. Fortunately, water and hydrogen is abundant enough that it will save us when there is a crisis in oil production that will eventually run out. The hydrogen gas engine has been patented for some time but that is another story of man's ingenuity. Take out a map of the United States and check the major large population areas. You will find several facts. One most of the major urban areas called "megalopolises" are located near large water bodies. For example, the North East Coast megalopolis includes Richmond, Washington, Baltimore, Dover, Ocean City, Rehoboth, Wilmington, Philadelphia, New York, Providence, and Boston. The Great Lakes megalopolis includes Toronto, Buffalo, Niagara, Erie, Cleveland, Detroit, Windsor, Toledo, South Bend, Gary, Chicago, Milwaukee, and Green Bay. The California and Northwest Coast megalopolis includes San Diego, Los Angeles, San Francisco, Portland, and Seattle. The Gulf Coast megalopolis includes Brownsville, Houston, New Orleans, Pensacola, Tallahassee, St. Petersburg, Fort Meyers, and Key West. The South East Coast megalopolis includes Norfolk, Virginia Beach, Kitty Hawk, Myrtle Beach, Beaufort, Savannah, Jacksonville, St Augustine, Daytona Beach, Ft. Lauderdale, and Miami.

The rivers and tributaries of these zones of the United States are very conducive to building and water treatment plants by mankind. In this regard, the ecosystem is dependent on man to develop the land accordingly using environmental friendly laws. The Chesapeake Bay watershed region runs from Pennsylvania down to Virginia through 4 states. The point here is that natural life burgeons in this region due to the single fact that there is an abundance of water, good ole H_2O. Transportation by water is also available to all these locations. This was the major transportation mode before air, train, and automobile over 100 years ago. This was a major reason for large cities developing in these areas. Figure 6 shows the states with the largest water areas. It is no coincidence that these states have cities are highly populated and near a water supply. The coastal megalopolis's lie mainly in these states listed. A water supply has great implications for sewage drainage and health reasons as well as drinking supply. Biologically, man can not survive without water. Minnesota is the only state listed that is not bordering on a major body of water and is known as the "Land of 10,000 Lakes" and some quite good fishing. Alaska is so vast we need to urbanize more there. The other 8 states are all in coastal regions housing the cities that are the coastal megalopolis of the 21st century and beyond.

Figure 6. Top 10 States Water Areas

Rank	State	Water Area
1	Alaska	91,316
2	Michigan	39,912
3	Florida	11,827
4	Wisconsin	11,187
5	Louisiana	8,277
6	California	7,736
7	New York	7,342
8	Minnesota	7,328
9	Texas	6,783
10	North Carolina	5,107

The election of 2004 proved a point when one looks at the blue states and the red states. The northern blue states and west coast blue states indicated that major northern cities voted for John F Kerry and democrats. The southern red states voted primarily for George W. Bush. I think of how people in major cities are more open minded than those in rural areas who own land and large amounts of property. There is definitely a pattern of major northern urban areas voting for Kerry. The map looks interestingly like a map of the civil war states and how they were aligned based on the slavery issue. I would think one might analyze the spectrum of liberal to conservative voters living in those areas also. Maybe we have much more liberal democrats in cities and the areas that voted for Kerry and conservative republicans in the southern states. The coastal megalopolises are centers for future voters in all elections as major urban areas. The southern coastal megalopolis areas have especially gained population since 1960 as people seek sun belt areas of warm climate for health reasons or economic reasons.

24. Quality of Life Measures and Statistics

In most scientific studies this would be a fair place to start a discussion on comparing cities statistically to come up with a measure of quality using various quantities such as rates and census data. Data such as education level, crime rates, birth rates, death rates, income level, accident fatal and injury rates, life expectancy, years of marriage, sex, health care availability, number of doctors, number of schools, amount of technology, transportation alternatives, miles of highways, access to public transportation, number of sidewalks, number of teachers, number of lawyers, and other demographics measures of quality can be factor weighted for the various cities of the United States and this was done in one study by the author in undergraduate school in a 300 level research methods course. The result of the paper study was that the cities of Dallas and Seattle had the highest quality of life in 10 major urban cities that were compared in the study. The problem with the study was that the author was trying to access where one would live the best life in America in 1978. The 1970 census data was used from the University of Maryland library to determine the quantitative scores for each city. It is much easier to see the limitations of this approach all these years later. Quality of life is a general term and it may vary between people based on their personal likes and dislikes. However, general indicators of quality of life have not changed since the 1970 census data. The study should have included smaller and medium size cities in less urban areas. The data are also very difficult to obtain across countries. Cities in various countries may have very difference quality of life measures. The author has learned many new aspects of living in foreign countries and a new study needs to include all the cities of the world over 100,000 and all the cities under 100,000 to be a more valid study. Data for this will be obtained from the Time Almanac for 2005 and Encyclopedia Britannica. Then one could also do the study area as the various countries of the world and include any city in that country. The hypothesis of the study will be that quality of life is better in the developed countries of the world. Smaller urban areas located outside the SMSA may also have a higher quality of life. Satellite cities may have a better quality of life even with longer commutes to work. Urban and rural areas in warm climates may also have better health year round than urban and rural areas in colder climates.

The ideal study would use the following factors in some sort of weighted factoring to determine quality of life in various world cities:

Average Education level

Average Income Level

Birth Rate

Death Rate

Fatal Accident Rate

Life Expectancy

Weather

Average Commute Time to Work Time

Traffic Congestion

Average Home Prices

Low Crime Rate

High Rate of Home Ownership

We can use SAS, SPSS, or just a regular RDBMS database or spreadsheets can cross tabulate the data into meaningful statistics to assist us in finding the best quality city to live in. The Time Almanac has a category for cities with the lowest crime rate or safest cities over 75,000 people shown in figure 7 below.

Figure 7. Safest and Most Dangerous Cities in the United States

(75,000 population and over)

Top 10 Safest		Top 10 Most Dangerous	
Rank	City	Rank	City
1	Amherst , NY	1	St Louis, Mo.
2	Brick Township, NJ	2	Detroit, Mich.
3	Newton, Mass	3	Atlanta, Ga.
4	Thousand Oaks, Cal.	4	Gary, Ind.
5	Sunnyvale, Cal.	5	Baltimore, Md.
6	Cary, N.C.	6	Camden, NJ
7	Orem, Utah	7	Compton, Cal.
8	Clarkstown, N.Y.	8	Flint, Mich.
9	Mission Viejo, Cal.	9	Tampa, Fl
10	Lake Forest, Cal.	10	Jackson, Miss

Source: Time Almanac 2005, pp 205

Figure 8. Crime Rate of Top 10 Cities in Population

City	Crime Index*
New York, N.Y.	3,286.7
Los Angeles, Cal.	5,029.3
Chicago, Ill	**
Houston, Tx	7,106.6
Philadelphia, Pa.	6,183.1
Phoenix, Az	7,681.8
San Diego, Cal.	4,048.0
Dallas, Tx	9,132.1
San Antonio, Tx	8,243.3
Las Vegas, Nv.	4,524.2

Source: Time Almanac p. 388

* Crime Index = Murder+Forcible Rapes+ Robbery+Aggrevated

Assault+Burglary+Larceny+Motor Vehicle Theft

**Some indicators not available for total crime index

The problem with such a study is that the data would not be in like units such as price, miles, meters, income dollars, pounds, rubbles, pesos, lira, etc. This is a problem the U.S. congress has been unable to solve since the 1800's but they would never tell you that. They are too weak because of special interests and have determined the metric system conversion will cost too much to impose a standardized set of metric units on the United States due to special interest groups rather than thinking of the overall good of the world. So we work in the dual system with our own units and metric units. There may also be some identify lost if the United States gives up it's systems of measurement and Great Britain does not go to the Eurodollar. They are already on the metric system. Maybe Congress does not want to appear to be following the rest of the world. The lack of conversion also costs us money because of the non-standardization of products we export to other countries. Our products would be more acceptable overseas if we mandated metric system usage in all systems and we would better climb out of our recent trade deficits. Foreign manufacturers are still using the metric system for US imports so we stay a dual system nation.

Part II. Cities of the World

This part of the book is dedicated to the major cities of the world and describes them in enough detail to draw generalities and introduce the reader to their major historic components using maps.

25. Cairo, Egypt

Cairo is the city of ancient Egypt. The pyramids are nearby in Giza. Cairo is located along the Nile River. A trip to Cairo is said to enlighten the sense of old world knowledge in a person.

Figure 9. Cairo

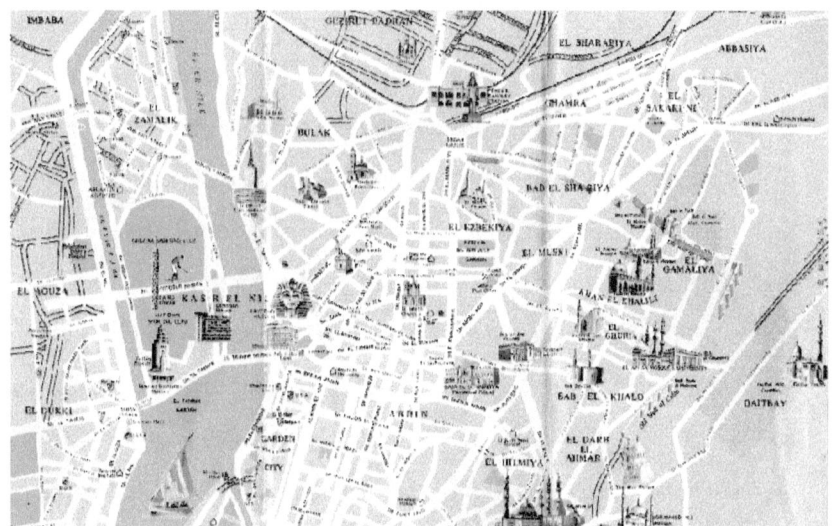

26. Baghdad

Baghdad has been the site of intensive fighting during the Iraqi war. The
study of the urban planning elements of the city could well be used in urban
warfare combat. Certainly if one knows the geography and layout of a city it
helps in the fighting of insurgents who may not know all the geography and
history of the city. Baghdad's history is as old as the Old Testament in the Bible.
The ancient culture of King Nebachanezzar dominated the Jewish people of that
time enslaving them. Ancient Babylon was located in modern day Iraq. The
Tower of Babel was created in Babylon.

Often the region of Mesopotamia is called the "cradle of civilization" where
mankind first built large cities. With a culture as ancient as Babylon, all of the
world needs to pay attention to the designs and architectures of Baghdad.
Building techniques there are centuries old. A map shows the structure of the
city of Baghdad and the major highway network and buildings. The Euphrates
river runs through Baghdad. TV videos are giving Americans a front line view if
combat action in Iraq. Where the urban environment causes problems in combat
is in the density of the buildings and the visual range reductions of warfare inside
the city. During world war II there was fighting in cities in Europe that simulated
urban warfare of Iraq. The rebuilding cost after such campaigns is significantly
higher than non urban warfare or jungle warfare as we fought in Vietnam. In fact,
neutron bombs can kill the people and leave the buildings standing for later use

by an invading force. This is a powerful weapon that preserves the infrastructure of the city. The contracts to private builders have been numerous during the Iraqi conflict and have been a topic of concern. The fact is that sometimes private contractors can do some things better than the military and construction work may be one of them. The Army Corps of Engineers are very good but they would be stretched thin without the private contractor help rebuilding the civil engineering projects that the citizens of Iraq have been complaining about like water, sewer, and electricity. Geographically, the city of Baghdad has a central business district, inner suburbs, and outer suburbs like many other cities of the world. The river splits the downtown area. Hoyt's concentric circle theory also applies to the city. The city is located in the center of Iraq with major cities to the south and north. The southern cities have the seaports on the Persian Gulf. The northern cities are near the mountains where the Kurdish people live. The country is Muslim which means they have Mosques instead of churches as we know them in the United States. They pray 5 times a day on Fridays. The mosques do not have seats like our churches do since they pray on their knees.

Figure 10. Baghdad

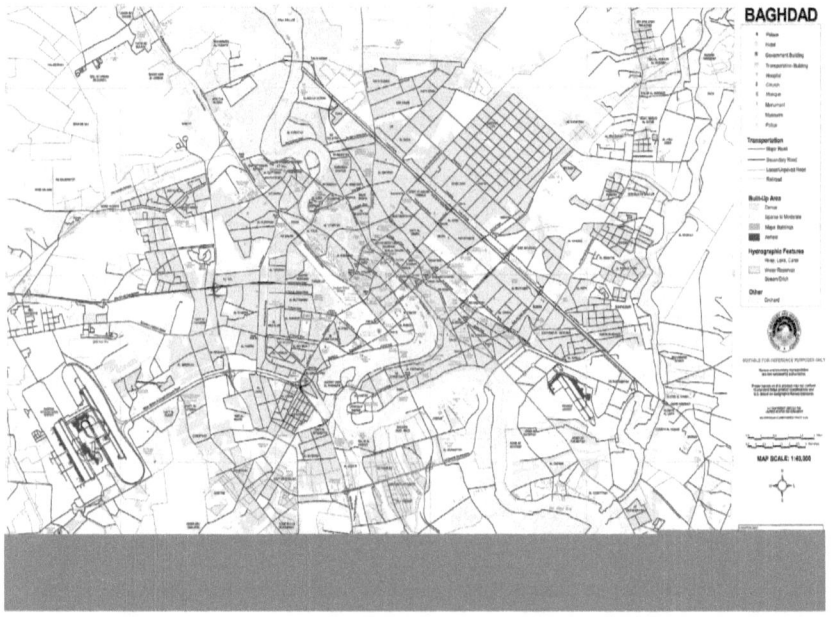

27. New Delhi

New Delhi is the capital of India. India has many provinces with many

kings and lords with an ancient history all her own. Only in the last 100 years

has the Royal British Empire sought to make India her own. They have

succeeded in developing a new culture in India but they have not succeeded

in destroying ancient Indian customs. It is good that India should maintain her

identity. She was a world war II ally of America and Britain and today she has

a vast army of employees who are skilled engineers, technical people, and

sales people all over the world.

Figure 11. New Delhi.

28. Athens

Athens was the academic city of Greece. It was well known for it's academic studies and philosophers. This year, 2004 the Olympics revisited Athens. They originally started in Greece as a way to keep the population healthy. This is a gift to westerners. The philosophy from doctors of medicine that sports can help keep us healthy. Actually it was designed to keep the Greek soldiers healthy and game for anything they might encounter. Much of the western world today prescribes to the traditions of the Olympics and healthy living standards. The Parthenon is located in Athens on a Hill near the center of the city. I have flown over it many times in Microsoft's Flight Simulator. So much for visiting every major city personally. We saw a lot of Athens in the 2004 Olympic games between events. The weather is hot there in the summertime and perfect for outdoor sports. The city is located on the Mediterranean Sea and highly accessible to shipping and cargo trade.

Figure 12. Athens, Greece.

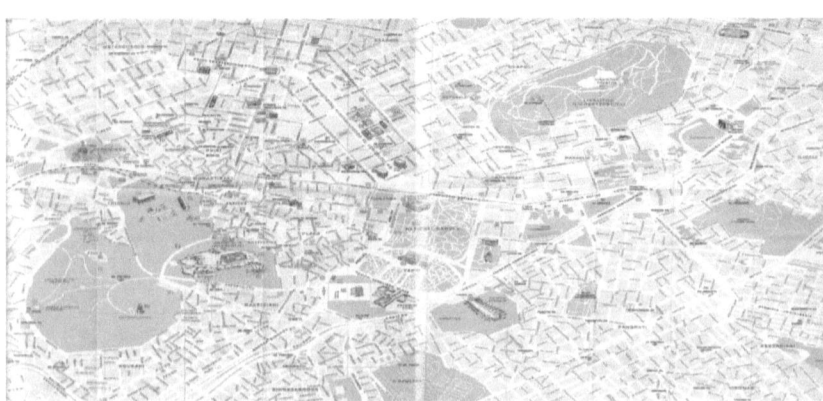

29. Troy

Troy is in modern day northeastern Turkey. It was sacked and rebuilt seven times in the ancient world. After the Greeks beat the Trojans in the Trojan War Aneas escaped to found Ancient Rome. A book was written about how that happened called the Aneid. Troy was ruled by King Priam and his two sons Hector and Paris in those days. Paris stole Helen of Troy from King Menelaus and that started the Trojan War. All the Greeks came to Troy camped on the shores and had a ten year war with the Trojans to get Helen back for their King. Her beauty was well documented. Today we say she had the Face that set sail a thousand Greek Ships. The book called the Iliad is about the Trojan war where the Greeks sacked Troy. Today we get the term Trojan Horse in computer security and other security terms from the Trojan Horse given to the city of Troy by the Greeks with hidden soldiers inside the belly. Once the gift horse (From the saying "Don't look a gift horse in the mouth or you might see soldiers") was inside the impregnable city walls of Troy the Greek soldiers came out and sacked the city. This is also where the term "Beware of Greeks Bearing gifts" came from. Imagine New York being sacked and rebuilt seven times. Hard to imagine.

30. Rome

Rome was the ancient center of the western world under rule of the Romans from early BC until 380 AD when the Goths and Visigoths sacked Rome. Rome was founded by Aneas who escaped from Troy when it was sacked during the Trojan war with the Greeks. Legend states that Rome was founded by Romulus and Remus who were raised by wolves but this is a folk tale. The Roman empire aspired to become the great republic that stretched across Africa, the Middle East and Europe into what was called the provinces. The Roman rulers brought technologies and engineering to the provinces. In fact, Roman engineering built London twice. Once before a fire and once after a fire destroyed London. London was an outpost of the Roman army. The Legions did not enter Scotland above Hadrian's wall. The Scots proved to be a hard opponent to the civilized Romans. The Romans built highways and fortifications that still remain in Britain. The Roman influence was so great that in the 1700's the brilliant professionals of Britain went back to Rome to study architecture and engineering. Rome was the first time a republican form of government with Senators was formed. It was good to be a Roman citizen. Slaves were captured from enemies of Rome. Rome was oppressive to nations it wished to conquer under the Ceasars whom were viewed as Gods. The Greek Gods were worshiped in Rome under new names. A new religion named Christianity was born in the Roman occupied city of Jerusalem when a roman governor named Pilate crucified Jesus

Christ. The country became the first Christian state under Constantine.

Gibbons book series chronicles the Fall and Decline of The Roman Empire

which coincides with the acceptance of Christianity and a dying of the rule of

dominance. Roman architecture can be seen in current day American

government buildings, museums, libraries, and other buildings.

Figure 13. Rome

31. Jerusalem

The capital city of Israel where Jesus was crucified and resurrected. The city is divided in to four quarters with various religious backgrounds. Muslim, Christian, Jewish, and a forth. The wailing wall is one of the greatest attractions from ancient times for the Jewish tradition and faith. The holy sepulcher where Jesus was buried is also an attraction for all. The city has been modernized since ancient times and I the focus of much of the free world and evangelists and other religious men and women. Many pilgrims make their way to Jerusalem to walk in the steps of Jesus during his crucifixion called the stations of the cross through the present day streets of Jerusalem. This is also done at Christian churches all over the world at Easter time. The Jerusalem Post is a modern day newspaper from Jerusalem available to most Americans online. The city was where the governor of Judea resided in biblical times, Pontius Pilate and King Herod, King of the Jews. The figure below show Jerusalem in ancient times of Jesus. Golgotha is where Jesus was crucified and The Garden of Gethsemane is where he was betrayed by Judas Iscariot. The pools were religiously important for cleansing.

Figure 14. Ancient Jerusalem

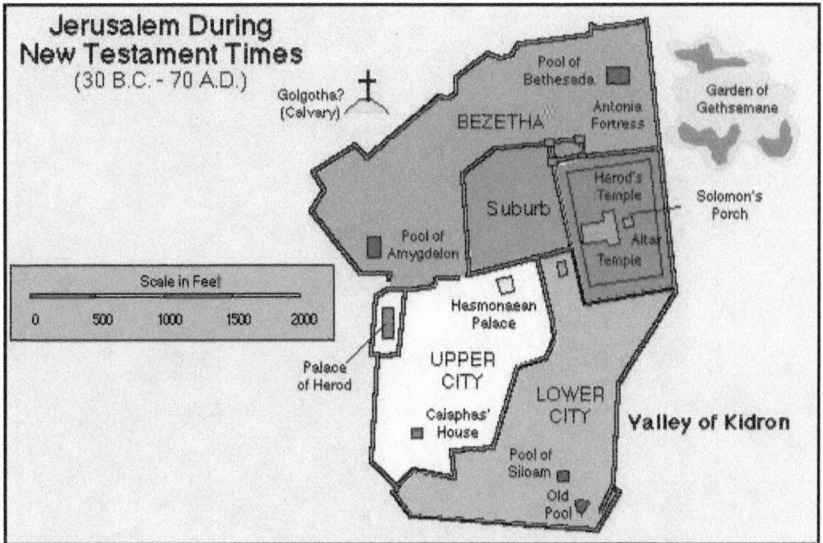

In the map of modern Jerusalem one can see the Old City. Hebrew University also takes a good portion of the land. No man's land is the disputed areas around the city suburbs. The city has grown a lot since ancient times.

Figure 15. Modern Jerusalem

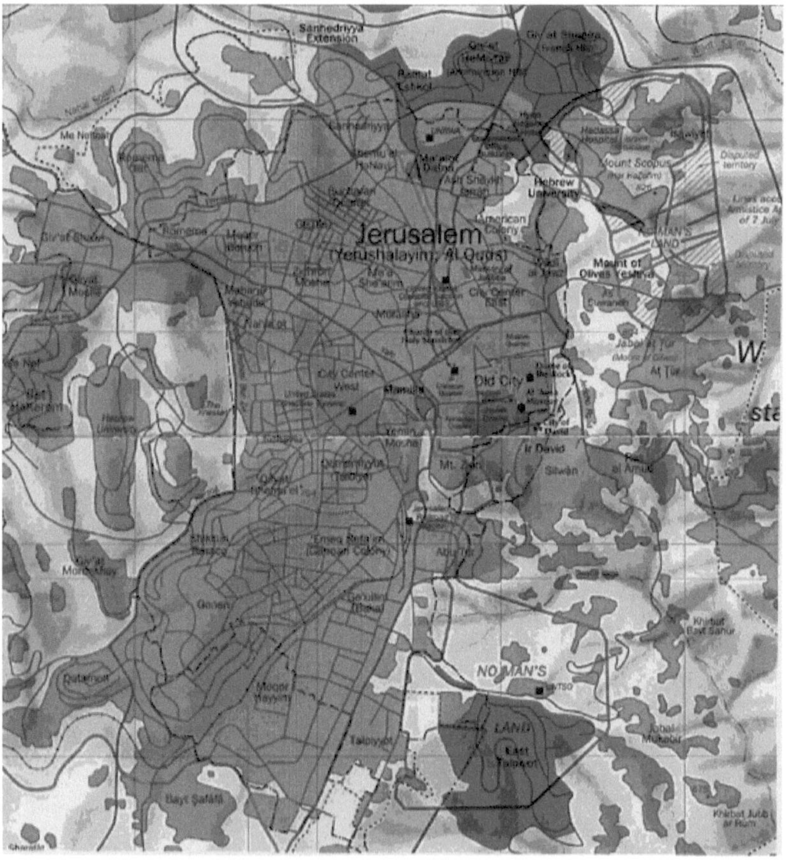

32. Berlin

Berlin was totally destroyed at the end of World War II. The city was then divided between the east and the west. The Berlin wall was the marker that divided the two halves of the city during the Cold War as shown in figure x below. The Russians controlled East Berlin and the Republic of Germany controlled West Berlin. With the falling of the Berlin wall in 1989 under President Reagan, the people had time to celebrate by taking small mementos of the rubble. This city was the major hub of Cold War activity between the East and West. Before the war Berlin was a center of German industry and culture. As Germany self destructed under Nazi dictatorship rule of Hitler it became the center of focus for much of the Allied drives through France and Belgium during the summer of 1944. The 1936 Olympics were held in Berlin and American Jesses Owens won a gold medal there in the 100 meter sprint downplaying Nazi racist supremacy. The Berlin Airlift was an attempt by the American US Air Force to airlift goods and supplies to East Berlin after the Cold War started in 1946 to help the people of East Berlin. East Germany represented East Berlin in the Olympics for many years after 1946 until the fall of Communism in 1989.

In medieval times Germany had many castles and nobles who protected the country. Lutheran Protestantism started in Germany under Martin Luther. He challenged the Catholic church and created a religion where every lay person could contribute. He wrote a book called 95 Theses that outlines his complaints to the Catholic Pope. These formed the protestant movement and a whole new view of the world separate from Catholicism.

In ancient times, Germanic tribes fought the Roman Empire for control of the Germanic province. Eventually, the Visigoths and Goths overran the Roman from the inside when they were invited to share the resources of the Empire. King Alaric was the reigning power of the Visigoths at that time.

Germany is also well known for the autobahn, a high speed highway running across the country, highly precise engineering products, German beer and wins, and lovely girls. The Germanic language is one of the more difficult languages to master but is required for local living.

Figure 16. East and West Berlin during the Cold War years (Pre 1989)

Figure 17. Berlin after World War II by Allied Occupation

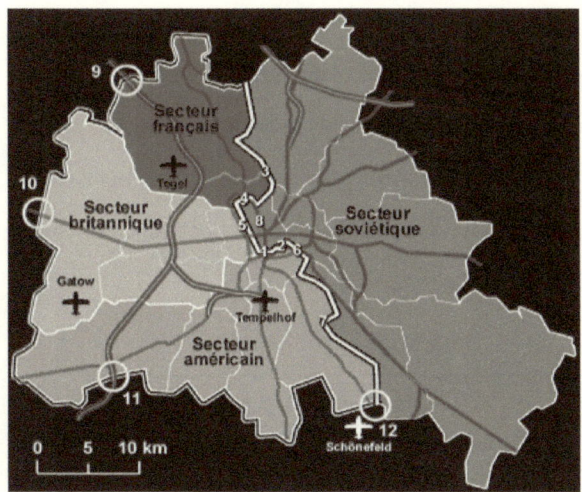

33. Warsaw

Warsaw is the capital of Poland. Poland is a country whose borders have changed many times over the centuries based on politics. During World War II, the Polish people suffered greatly at the hands of the Nazis. The Russians have been far less friendly since then. Poland has highly fertile agriculture and one of the oldest universities in the Eastern European theatre. Pope John Paul II was from Poland and was a priest at Our Lady of Cheztahova Catholic church. The majority of the country of Poland is Catholic.

Figure 18. Warsaw

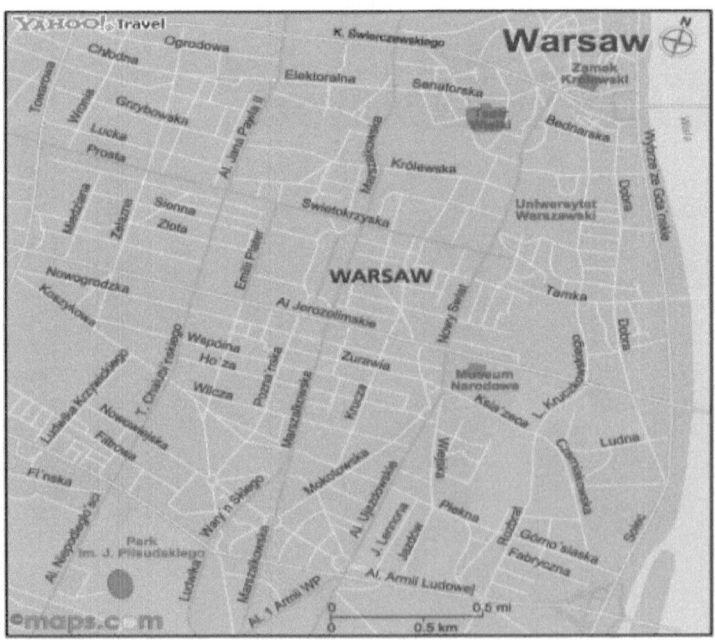

34. Paris

Paris is one of the nicest cities in the world to tour. The city is accessible
by air (Orley International Airport) or train from London through the Channel
Tunnel. The city has a beltway around it like some major American cities.
They have French food and theatre and all the amenities of the French
Lifestyle including wine and cheese and fruits. It was not very expensive to
stay in the downtown area near University of Paris. The subway system is
well developed in Paris and walking is encouraged around the downtown area
where there are plenty of parks and pathways. The Arc De Triumph has an
underground tunnel to get to it and you can go to the top of it. The Eiffel
Tower has a communications display at the top showing how the Germans
used it during world war II. Gustav Eiffel built the tower in the 1800's and
today it is a signature landmark of Paris. As you stroll down the avenues you
can see the similarities between Washington DC and Paris by their streets
design. The artists paint a lot of the cityscapes in Paris and buying some
artwork is highly recommended. Notre Dame Cathedral and St. Germaine's
are two of the finest churches in the world by architectural design. TV
stations are mostly in French. The Louvre has some of the most interesting
artwork and statues in the world on display to the public. Government
buildings look very similar to American government buildings. People there
are very friendly as tourism is one of the cities major sources of income. The
first phrase you need is "Parlez vous Englaise?" or "Do you speak English?".
I was surprised at the number of people who did speak English.

Figure 19. Paris

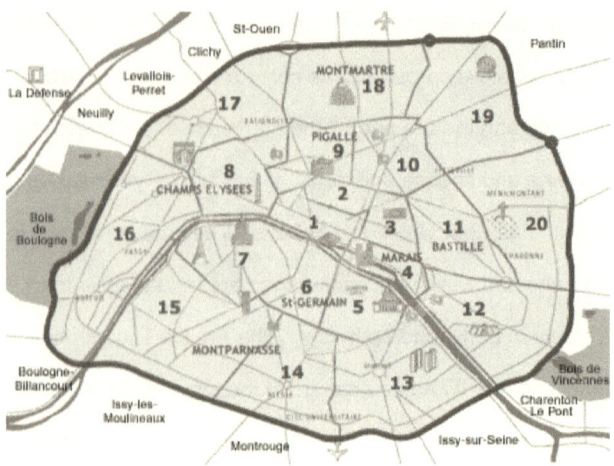

35. London

London was created in the 3rd century by the Romans who occupied
Britain. It burned down during a rat plague and was rebuilt when half the
city was lost. The blessing was that the rat plague also died in the fire. The
buildings were rebuilt in that downtown area and today thrives. The
current city is on the Thames River. It has one of the oldest underground
rail systems in the modern world. The city has contracted the American
Civil Engineering Society for many projects and these plaques are visible
in the tunnels.   Buckingham Place makes quite and impression also and
one may see the changing of the guards at noon. Trafalger Square is a
famous location in London with a statue of Lord Nelson the British Navy
Hero. The double deck buses are a joy to ride around the city. The tour
buses are wonderful for finding your way around for a cheap price. British
railway systems run from London to Paris through the Chunnel and are
also very reasonable compared to travel by other means.

Figure 20. London 1859

London has several international airports called Gatwick International
Airport and Heathrow International Airport. One of them handles local air
travel inside the United Kingdom exclusively to other Airports such as
Edinburgh and Glasgow Airports. Some of the historic places to see are The
Tower of London, Ceasar's statue, London Bridge, the HMS Belfast, St.
Paul's Cathedral, Trafalger Square, Shakespeare Theatre, The Tube (their
Subway), Oxford University, Bath, Parliament, and Westminster Abbey.
During the summer, the street vendors love to sell art and books of London
and England. This was very similar to sidewalk sales in Paris where
cityscape art dominated. It's also very nice to take in an English pub and
meet the people and have a nice meal. The cost of living in London is slightly

114

more expensive than New York and other American cities. The money used in the pound, shilling, half shilling, and pence. The banks are glad to post daily exchange rates with foreign currencies based on current market values. Electronic banking was not accessible to the United States banks. The city of London is pedestrian friendly but too large to walk alone. Many people travel by public transportation or private automobile on the left side of the highway.

Figure 21. London 2004

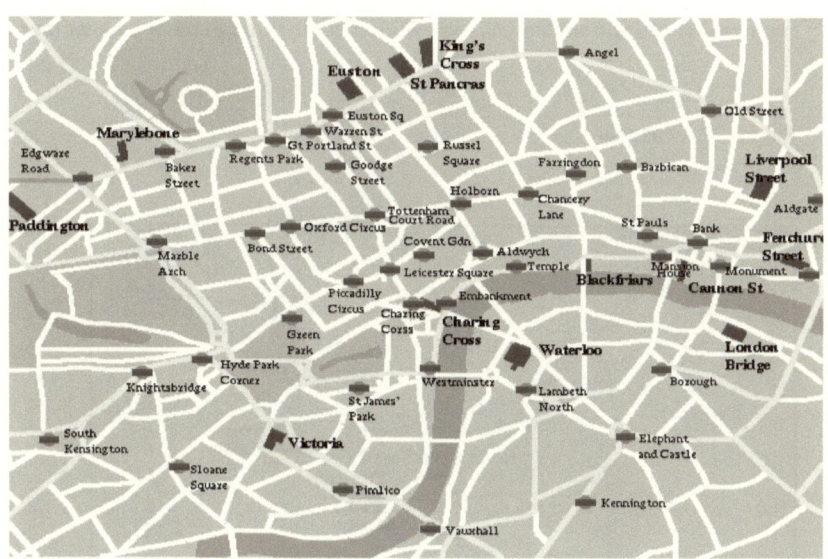

36. Glasgow

Glasgow is "no mean" city. It is located on the Forth of Clyde river. It is one of the industrial cities of Scotland that was bombed during World War II by the Germans to reduce Great Britain's industrial might. It is located on the west coast of Scotland just below the highlands. The city features a grand railway center in town and pedestrians are able to walk downtown in a traffic free environment. The University of Glasgow is the local university and is very good. Highway A77 runs from the Glasgow Airport south through other smaller towns.

Figure 22. Glasgow

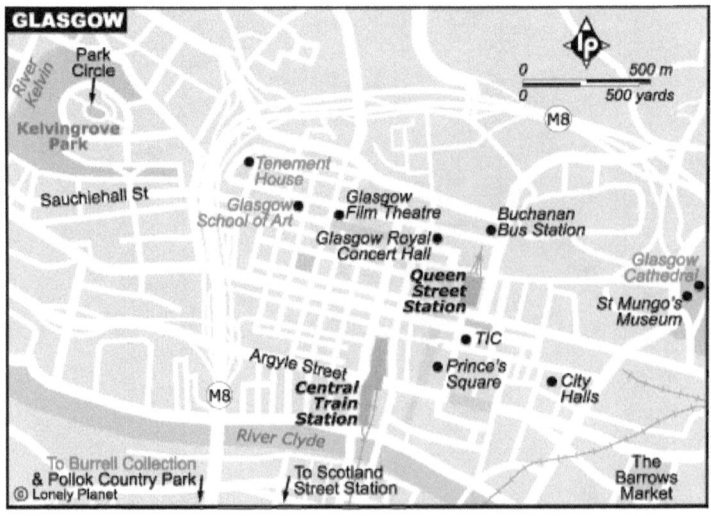

37. Edinburgh

Edinburgh castle and Royal Mile are the center of old town Edinburgh.
The railroad runs next to the castle through downtown. From the vantage
point of the castle one can see the ports and ships of Edinburgh. Edinburgh
University is world class among colleges and universities. If you are there in
the summer, you can see the summer festival and the Royal Military Tatoo.
The Stone of Scone is also on display. There are many historical tours
through the castle and Edinburgh. The city is located on the eastern coast of
Scotland in the lowlands. Economics are expensive in Edinburgh and
Scotland. The buildings are so old that most of them need sand blasting to
clean them. Prince street cuts right through the center of town. The castle is
on the highest hill in the center of town like something right out of a dream.
Edinburgh castle is one of the wonders of the free world to me. The castle is
truly a real life Walt Disney type of castle.   Maybe the better comparison is to
Mount Olympus of the ancient world. The high hill on which the castle stands
adds to the magnificence of the setting. Legends were made in the castle
and past history comes alive when you visit. The Scottish people are a proud
lot and this Castle is one of their crowning achievements in history along with
their vast understanding of economics, sciences, mathematics, engineering,
and medicine. Arthur Herman in his book "How the Scots Invented the
Modern World" describes the world of the 1700's during the period of
enlightenment when architects from Scotland such as Adams went to Rome

and brought back the best ideas of the ancient architecture. During this period of enlightenment the Scots also built many new cities and added to the beauty of Edinburgh. The transportation system in Edinburgh is primarily automobile now, but they handle pedestrian traffic by closing off entire streets during special events and at various localities in the city. The railway system runs straight by the castle in the downtown area from East to West toward Glasgow with a stop at the castle and Prince street. Railway travel is very inexpensive in all of Britain and Europe. This same train line leads down the east coast of Britain to London. The streets in the old town are traditional crosshatch pattern around the castle where the central business district is located. None of the buildings are skyscrapers of any huge size you might find in New York, but they are stately mid-sized buildings like you would find in Washington DC.

The churches are magnificently old and well preserved on the Royal Mile walk up the hill to the castle. From the castle you can see the harbor of Edinburgh and the battleships of the Royal Navy. Edinburgh has historically been a major port to Eastern Scotland. The suburbs are where most of the people live in Edinburgh and the homes are very modern without too many non conservative designs. They look rather dull in appearance from the outside but inside these homes can be very well decorated. Ancient Edinburgh is featured in the movie "Braveheart" during the scenes where the nobles are gathered. Edinburgh has always been a strategic city for

Scotland. The Germans tried to bomb Edinburgh during World War II and even thought of invasion. The people of Scotland knew they were targeted in Edinburgh because of it's industrial power and well being to the whole of Britain. The memorial chapel in the castle commemorate those Scots who gave their lives during all the wars. The chapel is also located at the best protected spot in the castle deep inside the double wall system. It seems almost impregnable but legend had it that the castle walls had been breeched. When you look over the walls and down the steep embankments you could never imagine a foe successfully attacking this castle. Edinburgh has modern highways as does the rest of Scotland and England only they call them motorways not interstates like in the United States.

Figure 23. Edinburgh

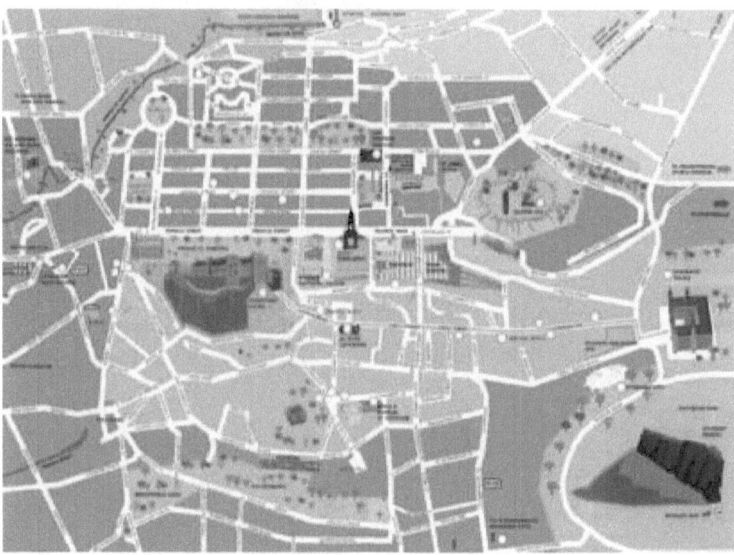

38. Madrid

Madrid is the capital city of Spain. There are bullfights there is the summer. Author Ernest Hemingway described Madrid in his ex-pat rioted views of the world in his book The Sun Also Rises. The city is traversed by avenues and boulevards that connect the major arteries and provide for business and economic development along the way.

Figure 24. Madrid

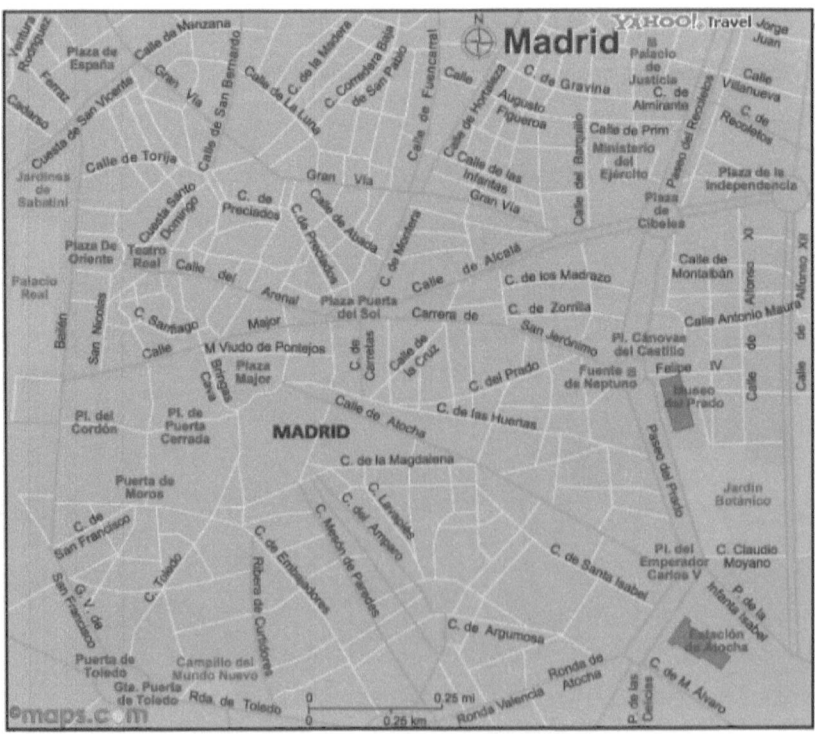

39. Boston

Mayor: Thomas Menino (until Jan. 2006)

2003 Census Population: 581,616 (23rd in US)

    Boston has many cultural museums and locations specific to the revolutionary period of American history. The USS Constitution is moored in Boston Harbor. Paul Revere made his famous ride in Boston. The old church is still there from which he saw the lanterns noting "one if by sea and two if by land'. Logan International Airport serves the region. MIT, Harvard and many other outstanding Universities are located in Boston near Cambridge and the Charles River. The world famous JFK School of Government is located at Harvard. Beacon Hill is one the more expensive areas of the city with high income and high status residents. Various groups of immigrants have settled in neighborhoods around the city. Herbert Gans wrote about the Italian West enders in his book, Urban Villagers. He described the invasion and succession of the suburbs of Boston by newcomers to America such as the Irish and Italians in the 1800's. The Fireman's museum in downtown Boston is an interesting site to visit. The Celtics, Patriots, Bruins, and Red Sox call Boston home.

Figure 25. Boston

40. New York

Mayor: Micheal R. Bloomberg (until Dec 2005)

Borough Presidents: Bronx – Adolfo Carrion

Brooklyn – Marty Markowitz

Manhattan – C Virginia Fields

Queens – Helen M. Marshall

Staten Island – James P Molinaro

2003 Population: 8,085,742 (1st in US)

New York is my father's city, specifically Brooklyn. The city was originally colonized by the Dutch and called New Amsterdam. The name was later changed to New York before the city became the capital of America. This name was never related to the English city of York. New York City is composed of 5 Boroughs plus the surrounding counties that form the metropolitan area. Here are the counties and their corresponding Burroughs:

Queens County - Queens

Richmond County – Staten Island

Kings County – Brooklyn

Bronx County – Bronx

New York County – Manhattan

Additionally there is the New York SMSA which includes the following:.

Long Island - Nassau and Suffolk Counties

Brooklyn was originally spelled Bruklyn – this is German spelling.
Brooklyn was the fourth largest city in the country at one time. Of course,
Manhattan is the central business districts of greater New York. New York
does apply Hoyt's concentric circle theory away from the Manhattan
epicenter. Manhattan is built on bedrock which is how it can support the large
skyscrapers. New York City is served by two airports – John F. Kennedy and
LaGuardia. Fiorello LaGuardia was the mayor of New York. The subway
was built in the late 1890's. Before there were bridges the ferry brought
people to Manhattan. It was Robert Moses the New York City urban planner
of the 1930's who built modern New York City. He built parks, major
highways, the United Nations, and revolutionized the way city planners
handled finances. He attend Yale and obtained a masters degree in Politics.
He once even ran for governor when he was so powerful that every major
project went through his office. He reported directly to the mayors. He was
the second in command. The book Powerbroker chronicles how he controlled
power in New York City politics for 44 years from the 1930-1960's.

Figure 26. Manhattan

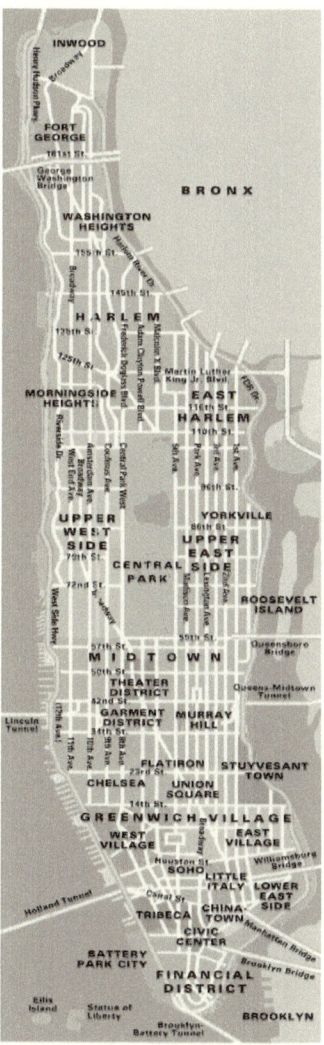

Figure 27. Brooklyn

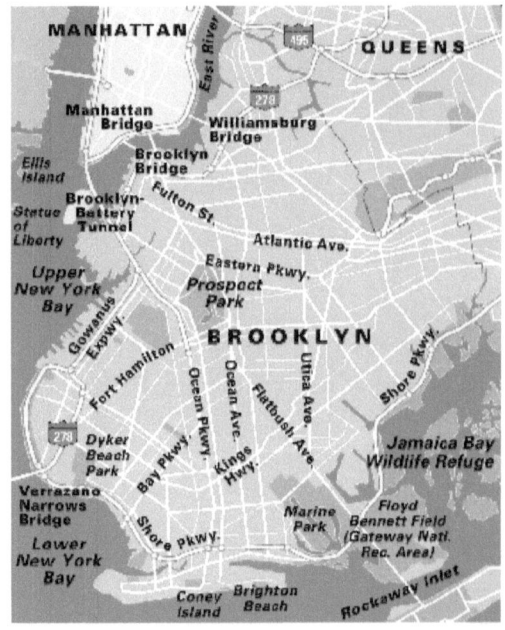

Figure 28. Bronx

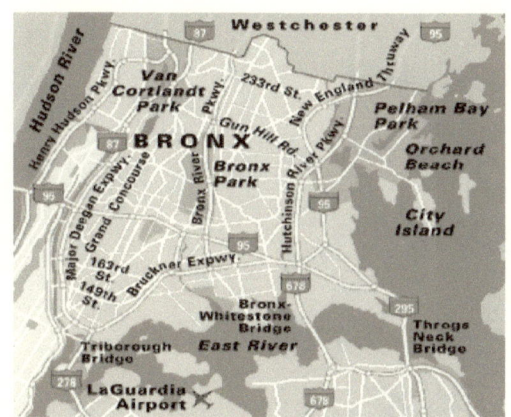

Figure 29. Queens

Figure 30. Staten Island

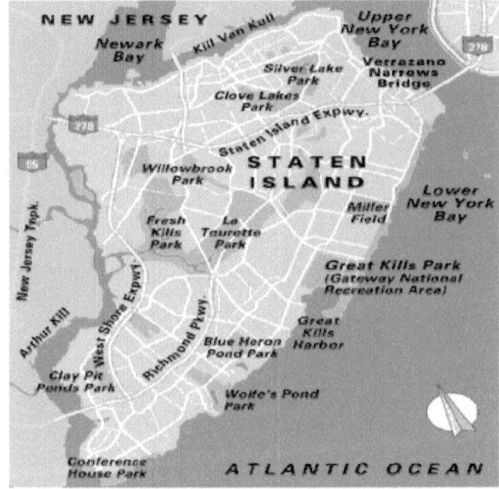

41. Philadelphia

Mayor: John F. Street (to Jan. 2008)

2003 Population: 1,479,339 (5th in US)

Philadelphia is an important baroque period city. As it stand today Philadelphia is two cities. The old capital of the United States (1776-1800) and the new city around the historic old city downtown. The old city has been preserved for future generation to enjoy. The historic city of Philadelphia is laid out very similar to Williamsburg. The museums and churches and freedom hall are all the same architecture as the Georgian style in Williamsburg. Some of the brickwork is original. The Franklin museum and docks distinguish Philadelphia as a unique port on the Delaware River. A navy town, Philadelphia has a history of shipping trades. The city was at the center of the industrial revolution.

Figure 31. Philadelphia

42. Washington DC

Date of Municipal Incorporation: Feb 21, 1871

Mayor : Anthony Williams

Motto: Justitia omnibus (Justice to All)

Flower: American Beauty Rose   Tree: Scarlet Oak

2003 population: 563,384 (25th in US)

Washington DC is a baroque period city built based on the design of Paris, France. Pierre L' Enfant and Benjamin Bannecker designed the city with the city in the baroque style using some of Williamsburg's characteristics. In fact, Washington DC and Williamsburg were only a horse carriage days ride away from each other. Washington was not always the capital of the United States. New York and Philadelphia and Annapolis were all the capital city of America at various times before Washington DC. Washington DC can not have any skyscrapers or buildings taller than 10 stories high. There is no building taller than the Washington Monument obelisk. The fact that Washington is a federal city makes it subject to federal finances and federal subjugation. The city has been trying to get home rule with two senators and representation in Congress for many years. Taxation without representation signs are everywhere. The city has a CDB downtown on K street. The pattern of streets is a crosshatch with avenues in a wagon wheel pattern. This was done specifically to prevent invaders from taking over the city. Gun emplacements were put in circles on the avenues to prevent an attack on

Washington such as happened in the war of 1812 when the White House was burned down by the British Troops and President Madison had to flee the city. By the civil war actual fortifications were built around the city to defend her from attack by the south. The Army of the Potomac was the federal army of the United States and the Army Corps of Civil Engineers built the city defenses. They also built the Washington Monument and the tidal basin as it is known today according to the History Channel. The area of the federal monuments was all imported from the dredging of the Potomac to make deeper channels for boats and ships. The Army Corps of Engineers has authority over all the waterways in America, not the Coast Guard or the Navy as one might think. One of the reasons in their expertise in bridge building and bridge demolitions is legendary as exemplified during World War II. The corps of engineers also presides over the drinking water supply in Washington ensuring it's safety. The cultural events are magnificent in Washington DC with a combination of museums and special events. The people are truly from all over the world and many are diplomats and dignitaries. There is a Chinatown area near the FBI headquarters building featuring many Chinese restaurants like Tony's. The federal triangle stop on the metro rail will let you walk around the federal museums area and Mall without worry about parking vehicles. On Saturdays and Sundays parking is easy in this section of Washington near the closed government buildings. Reagan National Airport services Washington DC. You might see Marine One fly over to the White House to pick up the President from Quantico

Marines Corps Base. The economy of Washington is based on government and services industries. There is no real manufacturing base in Washington which is very unlike it's neighbor to the north Baltimore. The Washington Marina is a small size boat dock on the Potomac river. The Anacostia River runs through the eastern part of the city and the less wealthy area of southeast Washington. Northwest Washington is the wealthy part of the town. This conforms the idea that most wealthy suburbs are located to the north side of most cities in the United States as a generalization. Hoyt's concentric circles of development can be seen in Washington as one travels from the White House as the CBD out to the suburbs. In fact, many a military study has been done using this as ground zero for a nuclear blast and counting the circle miles away from the epicenter of a fictional blast at the White House to determine who might be spared in the event of such an attack. A good book on the matter was Nuclear War What's In it For You? After reading this book I discovered that congress has a plan to escape from Washington to restart a new society should this event occur. My question is who will be left to rule? Or live? They might find it hard doing everything they ask of others themselves. But in strict survival terms, they have it figured out even with a new bomb shelter under the capital building that doubles as a visitor center. There are underground tunnels all over Washington DC from the government buildings to protect the federal employees also as a last resort. These tunnels evolved over time as ways to move people under the streets to other buildings to do business.

Figure 32. Washington DC Federal Mall Downtown

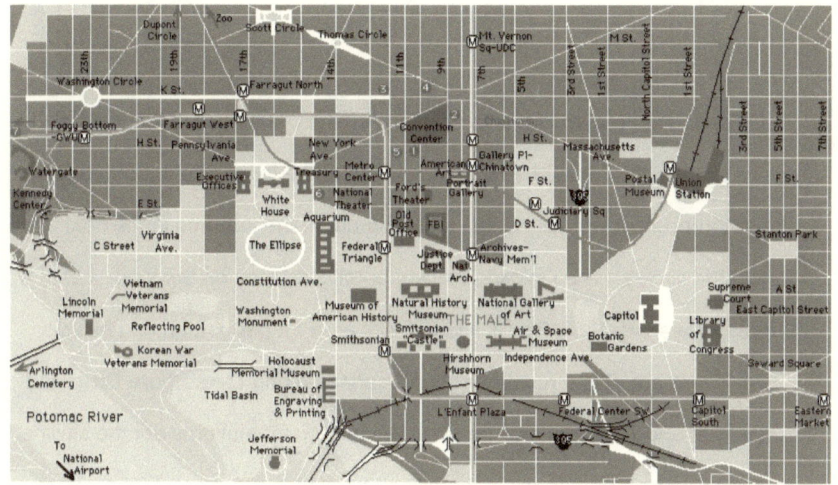

43. Chicago

Mayor: Richard M. Daley

2000 Census Population: 2,896,016 (3rd in US)

Chicago is one of my favorite cities in America. The downtown central business district is located on Lake Michigan in the loop. The L train runs straight through downtown to this area and out to the suburbs in the next concentric circles. The Chicago stock market is located on the loop. Lake Shore Drive along Navy Pier is a wonderful area for pedestrian and bicycle traffic. There are also a number of boat marinas along Lake Shore Drive and rental boat docks. Expensive skyscrapers and condominiums dot the skyline downtown. The Sears tower is the tallest building in Chicago. I-94 runs through the city north towards Wisconsin. The subway runs between the I-94 highway on long stretches of south Chicago inner suburbs near Comisky Park home of the Whitesox. Michigan Street at Christmas is an enlightening experience that every Christian will enjoy. The Chicago River runs through Chicago with river boats running daily. The Christmas lights are truly delightful for the children. The Urban Institute is located at the University of Chicago. Northwestern University Traffic Safety Institute is located in Evanston just north of the city. Chicago has a lot of railroads bring in the livestock from western ranches. Chicago also has a downtown area that is far larger than most cities. Chicago is largely a democratic city. Mayor Daily is a second generation mayor of Chicago. Chicago is the second largest city in America behind New York. The airports serving Chicago are Meigs Field and O'Hare International Airport. Flights out of O'Hare go all over the world. The Bears, Bulls, White sox, Cubs, and Black hawks are the major sports teams in Chicago and keep fans surprised. Wrigley field and Soldier field are both historic sports centers in Chicago. The great cities of America usually have two sports teams in some selected sports because the population of the city can support both teams. Chicago is this city in the Midwest. Chicago is a gateway to the northern Midwest cities of Milwaukee, Green Bay, Minneapolis St Paul., and Sioux City. Chicago universities include Northwestern, University of Chicago, University of Illinois, Northern Illinois, De Paul, and Notre Dame (in nearby South Bend). It is said a mid-western education is an honest hard earned degree. Chicago has been the site of many conventions and it's location in the middle of the country and near the great lakes makes it ideal for headquarters for corporations. The central location makes telecommunications costs cheaper to the rest of the United States. Chicago was well known for the gangs of the 1920's including Al Capone and Bugsy Malone gangs. Chicago public works are a very large operation with the number of patrons in the metropolitan area. Chicago is also a good pit stop place for travel to the far west states both by automobile and airplane.

Figure 33. Chicago Downtown

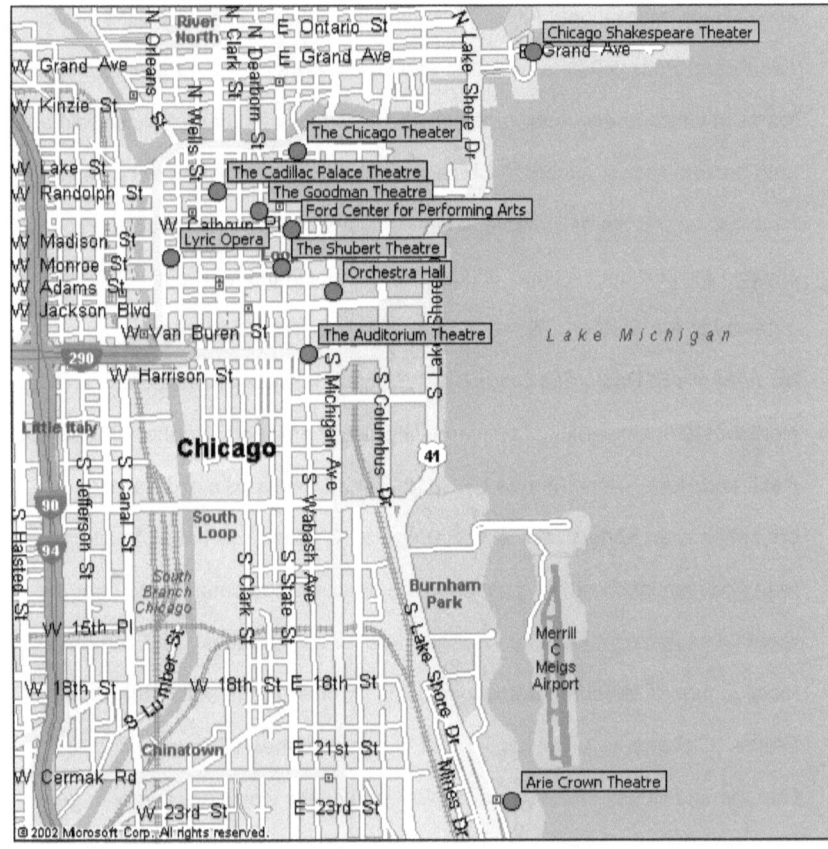

Figure 34. Chicago Metropolitan Area

44. Portland, Oregon

Mayor: Vera Katz (until Dec 2004)

2000 Census Population: 529,121 (28th in US)

The city of Portland won an award for the attention to bicycle traffic in the city.
The map of Portland's Bicycle paths shows the various routes one can take in
the downtown area by bicycle. The city is split by a river dividing east and west
Portland. There is a beltway around the downtown area central business district.
The streets are basic cross hatch patterns. Some research into the Portland City
master plan on internet proved that they have redone their master plans every
10-20 years since Portland was first developed as a city. The historical
progression of new suburban communities can be easily seen in these
successive master plans. It is harder to find these types of plans for cities in
other parts of the United States on internet.

Figure 35. Portland

Figure 36. Portland Bicycle Trail map

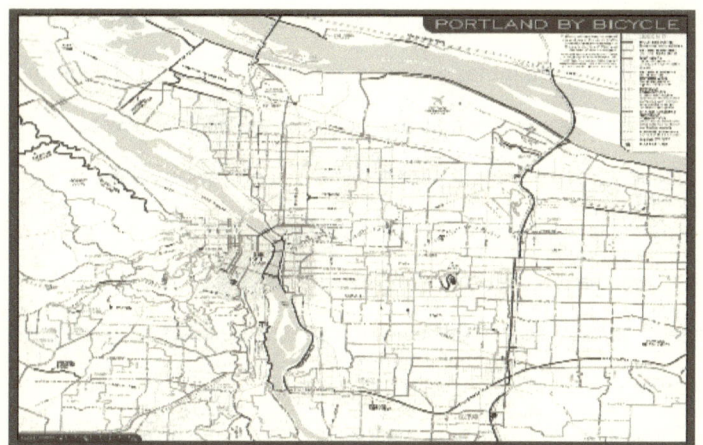

45. Tokyo

Tokyo is one of the largest cities in the world by population and population

density. Japan has 125 million people on a small island. At one time Tokyo

had 24 million people in the metropolitan area (1979). It would be wise for the

larger countries like America to study the characteristics of the culture of

Japan that allows the people to be so courteous and nice to each other in such small amounts of physical space. The population density in Tokyo is higher than any other city in the world yet they have the lowest crime rates. This is contrary to many countries who have extremely high rates of crime in their biggest cities. The exception appears to be New York City how has made in-roads in reducing crime in recent years and is now very safe. The culture of Japan allows them to be very high technology oriented and even world leaders in some technologies such as vehicles, electronics, computers, and high speed trains. The companies who manufacture these products are families who are very wealthy in Japan such as Hitachi, Suzuki, Fujitsu, Sony, Mitsubishi, Toyota, and others. The city is near to the holiest of places in Japan, Mount Fuji. Yokohama Bay is at the mouth of the city and the imperial Japanese Navy of World War II was based there. The map shows the train stations in Tokyo.

Figure 37. Tokyo

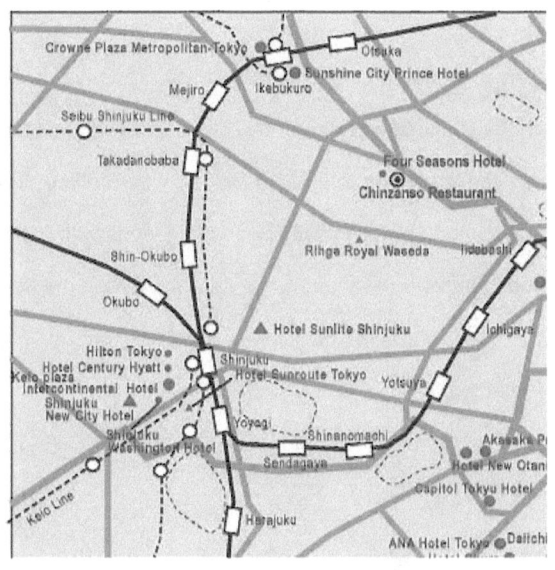

46. Hong Kong

Hong Kong is a commercial center in China. It was British property until 1999 when it was turned over to the Chinese government. The natural resources of the orient are the reason for the continued interest in China by the British. China and Hong Kong are both becoming an economic power in the world with inexpensive labor and goods and services that are very competitive in the free world.

Figure 38. Hong Kong.

For reference only. Map is not up to scale

47. Moscow

Moscow is the capital of Russia. It is also the location of the Kremlin and
other government buildings in the central business district downtown. The
interesting highways surround the city forming Hoyt's concentric rings as sen
in the map below. There are several parks including Gorky Park and others
in the downtown area to provide visual relief for pedestrians. The Moskva
river runs through the city and provides water to the citizens.

Figure 39. Moscow.

48. Oslo

Oslo is one of the great port cities of Norway. The map shows the significant

places to visit. Norway and the Vikings had the historic mastery of the sea and

they conquered many other countries from the sea including France, Scotland,

Greenland, and Nova Scotia. There was a map at one time that proved to be

fraudulent but stated that the Vikings discovered America long before Columbus

by traversing the North Atlantic. This would have been incredibly difficult but was

highly possible.

Figure 40. Oslo.

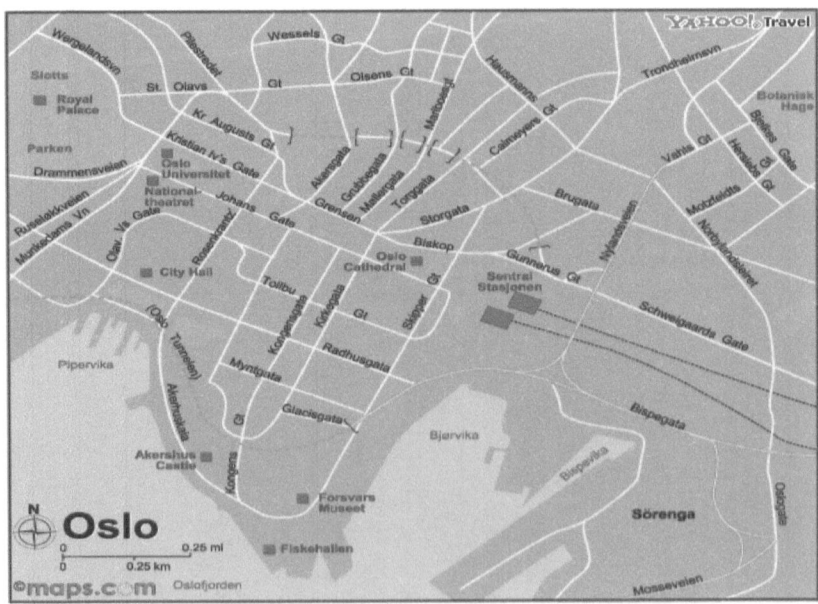

49. Mexico City

Mexico City was the site of the 1968 Olympics and they were televised on TV. The city is one of the largest in Mexico. Traditionally, Mexico had alliances with Spain in past centuries as well as speaking a common language. The American Mexican war was staged from Mexico City by the federal government. Santa Anna was the famous general who fought Americans in the war of 1836. The downtown is a crosshatch pattern of roadways with various shops and businesses located along the way.

Figure 41. Mexico City

50. Orlando

Mayor: Buddy Dyer (suspended for election violations)

2000 Census Population: 185,951

Orlando is a very spacious city with a lot of land mass inside it's city limits. It's central location is conducive to traffic from all directions and tourism is one of it's big industries. Orlando is home to many retirees who have found the climate to extend their years. Orlando recently experienced several Hurricanes which was the first time in 44 years. The central business district is very small but nicely located in the downtown area. The Orlando Magic basketball team call Orlando home. Located on Interstate 4, Orlando is a thriving new city with home prices in the upper regions. Orlando International Airport serves the city. Walt Disney World is a main attraction for children in Orlando. Orlando is a boater's and fisherman's paradise with the many lakes and oceans nearby. Traffic jams on I-4 are more frequent than in past years as congestion has hit the Florida interstate in recent years during rush hours.

Figure 42. Orlando

51. Seattle

Mayor: Greg Nickels (until Dec 2005)

2000 Census Population: 563,374 (24th in US)

The City of Seattle is located on Interstate 5 in Washington. The city is a major seaport and the Navy has a base there. The Seattle Ferry runs across Puget Sound daily. Seattle has several good universities. The central business district is located next to Puget Sound and Lake Washington. Seattle is famous for the Space needle, Supersonics, Mariners, and Seahawks football teams. Seattle is also home to Boeing Aircraft who has large investments in the city. The city has always been involved in aviation.

Figure 43. Seattle

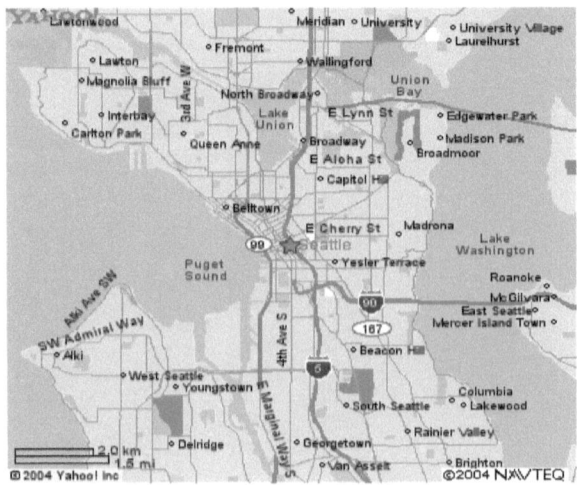

52. Baltimore

Mayor: Martin O'Malley (until Dec 2004)

2000 Census Population: 651,154 (18th in US)

Baltimore, Maryland was founded by Lord Baltimore when English

Catholics settled in Maryland. The city is a harbor port on the Chesapeake

Bay and a major sea faring city. Seaport industries dot the Baltimore harbor.

Ft McHenry is where Francis Scott Key wrote the Star Spangled Banner while

watching the fort being bombarded from a British ship in the war of 1812. The

Inner Harbor is the newer central business district of the city recreated in the

1980's to take on more visitors to this part of Harbor place. The USS

Constellation and USS Torsk are docked in Harbor place. Ships from visiting

nations often dock there. The Maryland Science center and Camden Yards

are close by the inner harbor as well as the Baltimore Convention center.

Pratt Street runs through the downtown just off the harbor. The World Trade

Center is located on Pratt Street and houses the Maryland Port Authority

among it's many tenants. The Baltimore Aquarium is located at the Inner

Harbor. Baltimore is primarily an industrial town. In 1904 there was a great

fire in downtown Baltimore that burned down huge parts of the city west of

Jones Falls. Fire companies from all over the battled the fire but could not

save the largest area of damage over many blocks of downtown. The city

was eventually rebuilt.

The Harbor Tunnel and McHenry Tunnels run through Baltimore along the bottom of the harbor. A beltway, I -695 was built around Baltimore in 1964. This coincided with the building of the Chesapeake Bay Bridge Tunnel that links Maryland to Virginia on the eastern shore. The Francis Scott Key bridge can be seen from the inner harbor when looking south. Baltimore is rich with heritage and culture of many parts of the world such as Polish, Italian (Little Italy), Irish, English, German, and Scottish. There is an Asian culture also in Maryland as well as Baltimore. The constitution of Maryland is based on the English Constitution. The local universities include Johns Hopkins, University of Maryland Medical, University of Baltimore, Loyola, and Towson to name a few. There is a thriving world class health care business in Maryland. Baltimore is also home to the Baltimore Orioles and Baltimore Ravens franchises. The city conforms to Hoyt's concentric circle theory outward from the inner harbor. The outer rings being composed primarily of residential suburban homes spreading into Baltimore County, Howard County, and Anne Arundel County. The Baltimore row house is a famous architectural structure located in the inner city housing near the CBD. I-95 cuts through the downtown area and provides major traffic flows from Washington DC and Philadelphia and Delaware in the north. Baltimore Polytech is one of the outstanding high schools for gifted high school engineering students in Baltimore and is very competitive. In recent years, Baltimore has lost total population to the suburbs due to quality of life issues. However, there are some very nice homes and parks in the inner city. Baltimore and Washington are more linked than ever before with both cities using BWI Airport and sharing the Baltimore sports teams. The traffic congestion in the entire Baltimore Washington region is ranked to be the second worst in the nation. This is due to high population densities in small land mass areas. Baltimore's city churches are well known on the East Coast. Baltimore was a place that helped run away slaves before and during the civil war. The underground was run through Baltimore by Harriett Tubman to get slaves to freedom in the north. By all counts Baltimore is very similar to a northern city that just happens to be located below the Mason-Dixon line. Baltimore is a primarily Democratic City in politics. The current mayor is Martin O'Malley, a democrat. A previous mayor, William Donald Schaefer, also a democrat, is now the Comptroller of the State of Maryland. The voting population of Baltimore and Maryland is primarily democratic as illustrated in the presidential election of 2004 when a majority of the state voted for John F Kerry for president and the electoral total of 10 votes for the states went to Senator Kerry. The larger populated counties in Maryland are all democratic, especially Prince George's County and Baltimore City. Politics in Baltimore affects all of Maryland. Sometimes Baltimore is mistaken for the capital of Maryland with many headquarters located in this large city, but Annapolis is the location of the seat of Maryland government. Baltimore is a very competitive city with other major cities such as New York, Philadelphia, Boston, Washington, and Pittsburgh. Baltimore has major league teams that have outstanding track records in every sport and the fans are a very proud

lot. The city is often viewed as having a large middle class and working class

background.

Figure 44. Baltimore

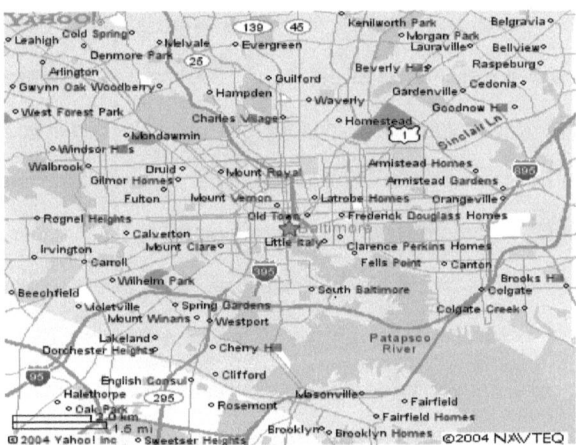

Figure 45. Baltimore Planning and Police Districts

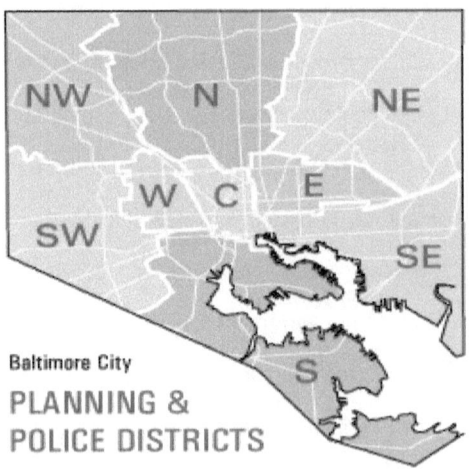

53. Annapolis

Annapolis is most famous for the Naval Academy and the current Maryland Statehouse and government. It is the capital of Maryland. It is the sailing capital of the United States due to situation on the Chesapeake Bay and the Naval Academy. The Naval Academy is the center of science and naval engineering and architecture programs training in the navy. Many other universities offer cheaper programs for a young person to become commissioned but few will challenge the individual like the Naval Academy discipline. Graduates are commissioned as ensigns in the US Navy and then the real lifelong learning begins a naval career.

Annapolis was a temporary capital of the United States and is where the Continental Congress met in emergency sessions when they could not meet in Philadelphia. Annapolis is also less famous as a slave port on the East Coast something which they do not play up. Tobacco was shipped out of Annapolis along with imports of slaves to work the tobacco fields in southern Maryland. Maryland is called the Free State but in actuality it was not always free for everyone. In fact, in the civil war the state was full of southern sympathizers although technically it was a northern state with mostly Union troops encamped here. John Paul Jones tomb is a visitors site inside the Naval Academy chapel. He was the founding father of the Naval Academy and strongly believed in training sailors at a site located on the bay. He was most famous for his words "I have not yet begun to fight" from the revolutionary war navy when he was being bombarded by British forces in a broadside attack.

Alumni Hall on campus is where the Naval Academy plays basketball against Patriot League foes. The annual Air Force vs. Navy game is usually an outstanding non-conference game played there for a nice price. It is rarely televised, but should be. It is less spectacular than the Navy – Army or Navy – Air Force football games which are never played in Annapolis due to lack of stadium seating for the demand by active duty and retired military families. I am proud to say that Navy now plays Maryland in football thanks to letters to the Naval Academy by yours truly and a change of heart by Naval Academy Management in what will be the best in state rivalry in Maryland college football. It took many letters and phone calls to convince them this would be a good idea for all concerned. Annapolis is home of the Maryland General Assembly and Maryland Senate. One can view the legislation in action during the days it is session by following the rules posted in the State House for visitors. The governor's mansion is also located in Annapolis. Many stores and shops dot the cityscape of Annapolis. The city is built in the baroque period style with a wagon wheel or streets around the statehouse and crosshatch streets downtown. The best way to negotiate the downtown area is by foot.

One can tour the St John's College campus grounds next to the Naval
Academy. They have a liberal arts program based on reading famous
literature. Annapolis Mall is located at the center of downtown near the
docks. The Naval Reserve Unit has a large building located next to the ship
docks. Annapolis from the water is a great way to see the city.

Figure 46. Annapolis

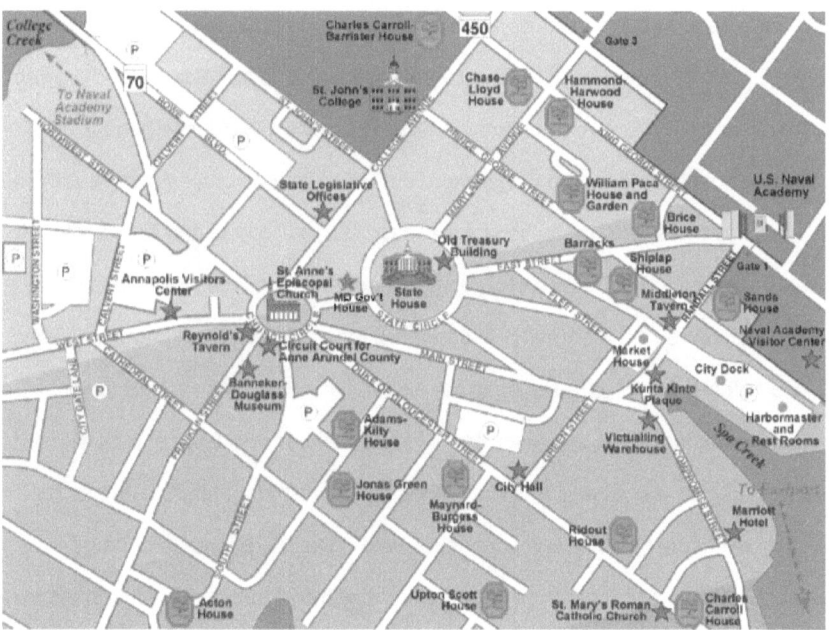

54. St. Mary's City

St. Mary's City is the historic first capital of Maryland when the Catholic and Episcopal people of England founded Maryland before moving north in the Chesapeake Bay to discover the city of Baltimore. Indeed, Lord Baltimore was Catholic as is a majority of Maryland to this day. The First Maryland Statehouse is a landmark that visitors can walk through when they visit the campus of St. Mary's College of Maryland. The college funded historic archeology land digs on campus in the 1976 year to uncover artifacts from the 1600's time period. This colony would have been thriving around the same time as Jamestown in Virginia. To this day there is a connection between St. Mary's College and Baltimore where families will send their children to school at the college to escape the city life of Baltimore. St. Mary's City is on the St. Mary's River and is today a thriving place to learn competitive sailing in the NCAA. They have won 10 championships in sailing against notable sailing schools like the Naval Academy and Coast Guard Academies. This is a distinguished accomplishment for a private college to achieve against the federal academies who build future leaders of naval service. St. Mary's City is also home to some of the oldest families in Maryland who are buried in the graveyard at St. Mary's Episcopal Church on the campus of St. Mary's College. The college has some outstanding liberal arts and sciences and is now counted as one of the best values in College Educations according to U.S. News and World Report. St. Mary's College is an honors college. It graduates many teachers into the teaching profession. It's location is ideal for undergraduate studies in history. They do not have an engineering program but do have a strong mathematics department.

The current site of St. Mary's City next to the campus is a registered national historic site and is open to the public for visitors interested in U.S. History. It is worth a day trip through southern Maryland to see the city and is not a lot of walking nor even as large a settlement as Williamsburg. In the summer there are plays and reenactments of the 1600's world of settlers. The Arc and Dove sailing ships are docked at St. Mary's City. These two ships brought settlers from England to America and are worth seeing in person. Both ships are rebuilt to original specifications.

55. Williamsburg

Colonial Williamsburg is one of those places you have to visit in the summertime when you can stroll around the town and talk with the actors who are playing colonial time parts. Parking is free. The English settled here in the 1600's and the buildings reflect the Baroque period pattern of urban development. In old Williamsburg you can catch a reenactment or two and see how they lived several centuries ago in Virginia. The Old Baptist Church was the first one in America and was for slaves. The Statehouse is styled in the Georgian style with red brick and is simply remarkably well preserved.

The stores and shops have original period items and plenty of history books on the events of that time period. A few short miles away and you have other small towns that are more contemporary. Williamsburg was settled during a time when America was in it's infancy. The downtown area is built around two avenues. The houses have the original carpentry on them as I was assured by an actor who worked on them and was living in one of them since he started his job in 1978. You could imagine the trip from Williamsburg to Washington in a horse drawn cart. Washington DC was also built in the baroque style as was Philadelphia and Annapolis. You can see the similarities if you visit these other cities. There is also a distinct similarity to St. Mary's City in Southern Maryland. The statehouse and other buildings are very closely similar in architecture of that period. Williamsburg reminded me of how far we have come as a country.

William and Mary College is located in downtown Williamsburg. It is one of the oldest in the nation. Madison, Monroe, and Jefferson attended college here. The area is very well maintained and the gardens are simply beautiful. The local church in downtown Williamsburg is Episcopal and still operational. This is not surprising since George Washington and many Virginians of that time period were Episcopal. Williamsburg is close to Yorktown battlefield where the Americans won their independence from the British when French ships supported them with artillery fire from the Atlantic, hemming in Cornwallis. The trip to Yorktown is worthwhile and the battlefield holds clues to the building of forts of that time period. The Newport News shipyards are not far away and provide the local jobs needed by the many people living there.

Figure 47. Williamsburg

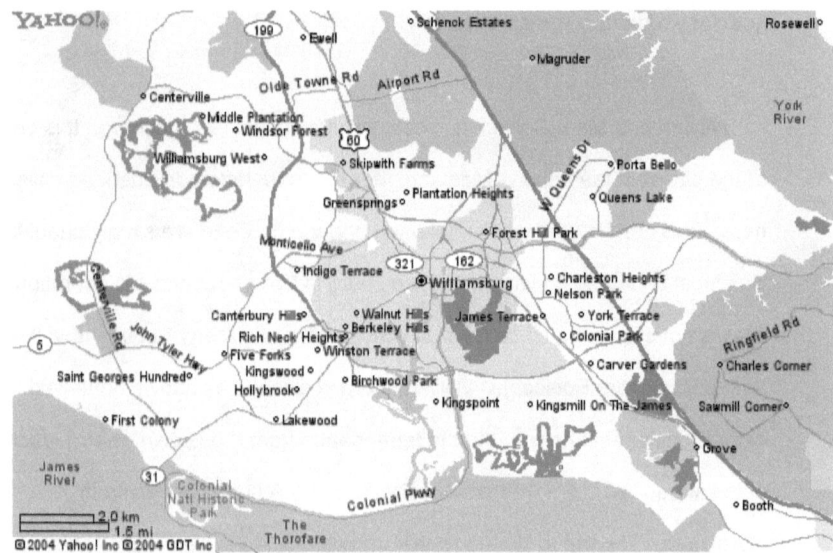

56. Toronto

Toronto was strikingly the most clean city I have visited in the world. The CN Tower is a must see attraction and the large dome stadium of the Toronto Blue Jays baseball team (which can be viewed from the CN Tower looking down). The city is located on Lake Ontario with a harbor and lakeshore property. Traveling there by vehicle one goes along the Queen Elizabeth Expressway. Toronto has it's own airport for air travel. The nearby golf courses are also worth a try for the golfer in the family. There were some French speaking natives in the city.

Figure 48. Toronto

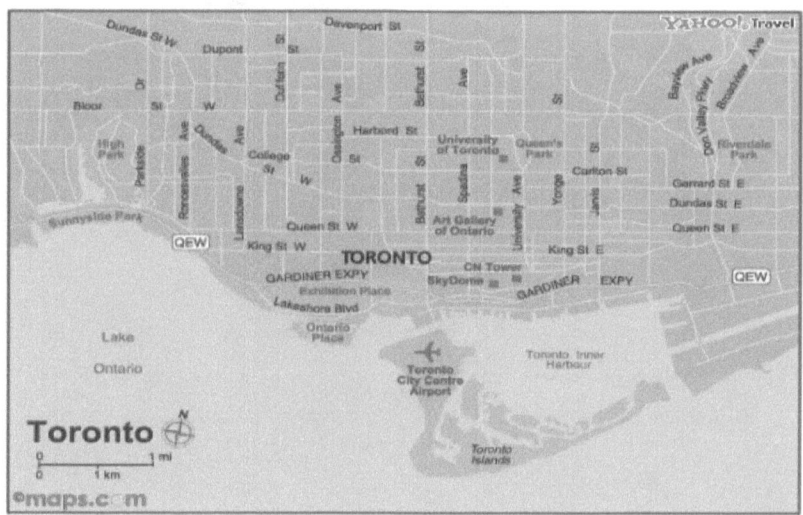

57. Quebec City

Quebec City has an upper and lower city outside the walls of Castle
Frantenac. The city is postured on the St. Lawrence Seaway and vessels
travel the seaway delivering goods. The old city is a lovely walking tour
through the inner walls. The local people speak French and consider
themselves to be a separate region of Canada and even sought
independence from Canada. The Notre Dame cathedral is a nice place to
visit also.

Figure 49. Quebec City

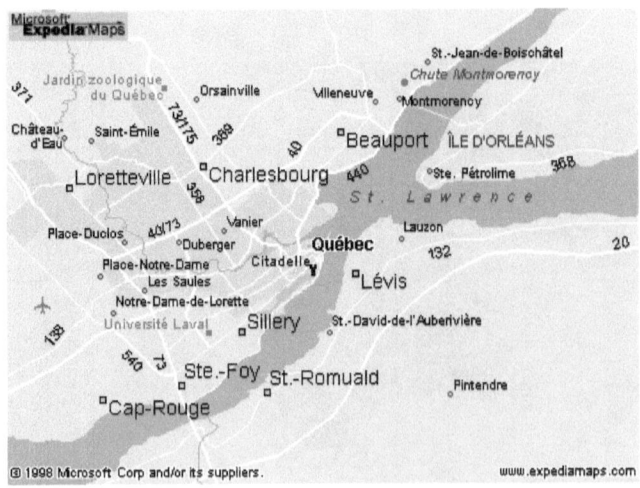

58. Bo-Rich

   Bo-Rich is the name given for the future Boston to Richmond megalopolis

as coined in the 1970's literature. Well the report is that in 2005 the name is

becoming ever more true. The census has created new terms for

Micropolitan areas as well that are filling in the areas between SMSA's –

Standard Metropolitan Statistical Areas. Despite the best efforts of

governments and councils private property owners and business owners

property rights are winning the land use battles and the lands which were

once farms are now developments and suburban in character. Smart growth

to quell the mass development effort simply is not working due to large

influxes of population and private money as people move up the economic

ladder. Land uses are regulated by county and local governments. Naisbitt

and others have predicted this for a long time in talking about regionalization

and regional thinking and actions of people in activity tracking and trip

generation. Movies and Hollywood are always exploring what the future

holds but these are guesses. The truth may be more exciting than we think

and less of a blessing. I look to two places on earth who have already

attacked the problems we will face in America in coming years when Bo-Rich

is a complete super SMSA or megalopolis – Japan and Great Britain.

America is fast approaching 285 million in population. The largest city is still

New York by far – right smack dab in the middle of good old Bo-Rich. Let us look at the former, Japan. They have 125 million people on a tiny island with $1/150^{th}$ the land mass of the United States. They have a great train transportation system (the Bullet Train) that we are trying to emulate in our urban areas here on the east coast with the Magnetic Levetation Train. The problem is we can not emulate the population density that Japan is experiencing. The amazing thing is that with 125 million people in close proximity, they are a very polite and crime free nation. They are also very stratified culturally and do not accept outsiders very well similar to most island nations. They are governing democracy with a multiparty system put in place after World War II.

The United States has invested a large sum of money to ensure the success and rebuilding of Japan after the war as we usually do. If Japan can handle this rapid development on such a small island then there is great hope fro Bo-Rich. Since we built modern Japan we can also use our own exported development processes here in America. We have already done this in auto manufacturing and TQM management sciences (Juran and Deming). The population of Bo-Rich includes the following cities:

Bo-Rich

Richmond

Washington DC

Baltimore

Wilmington

Philadelphia

New York

Providence

Hartford

Boston

The Southern California region has it's own megalopolis that has grown just as much since this prediction. The same with the Great Lakes Region.

California Coast

San Diego

San Jose

Los Angeles

San Francisco

Great Lakes Area

Cleveland

Detroit

Toledo

South Bend

Gary

Chicago

Milwaukee

Gulf of Mexico Region

Brownsville

Corpus Christi

Houston

New Orleans

Pensacola

Tampa / St. Petersburg

The pattern we see developing is man needs to be near water supplies. This has always been true since Mesopotamia. Ask farmers how they irrigate rural Kansas and how important this is to survival. The sewage systems in these areas are also well developed because they have been designed to handle the capacity of people living there. This is the real indicator of developing neighborhoods and cities. We can not build beyond the sewer capability.

59. Venice

Venice is located in northern Italy. It is a city with many canals and waterways. The Gondola is the vehicle of choice in the city. The map below shows the waterways through the city of Venice. Venice was the northern capital city in Italy and once held great prominence in the renaissance era.

Figure 50. Venice

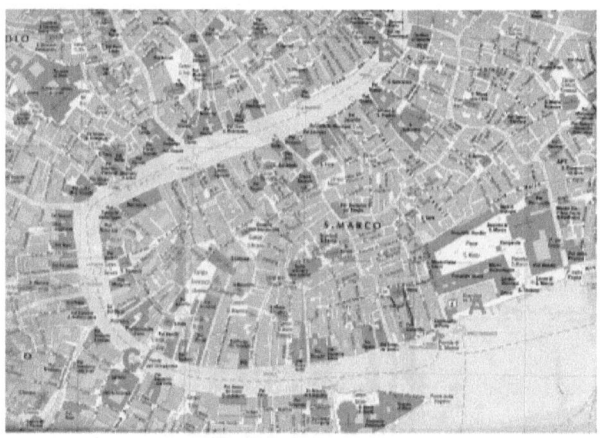

60. Montreal

Montreal means "The city on a hill". It is a large metropolis in French Canadian Quebec. The city is accessible by Route 15 from the USA – New York Interstate Route 87 north. Route 15 bypasses the city of Montreal and provides access to the mountainous ski resort region beyond Montreal. Many of the small towns around Montreal are named after Catholic Saints. The population is mostly Catholic and the churches are simply beautiful. They speak both French and English and you are best to ask if they speak English – Parlez vous Englaise?. The traffic signs are duel language coded near the border and in French the farther into Canada. The speed limit and distances are measured in kilometers. Prices on consumable goods are slightly higher than here in the United States.

Montreal is built next to a hill. The Central Business District is on a river and flat. Skyscrapers crown the downtown area. The Cathedral of Notre Dame is a great place to visit if you are nearby the downtown area. Olympic stadium houses the Montreal Expos baseball team in a state of the art facility. The winters are long and cold in Canada but winter sports are king. The Montreal Allouettes are the CFL team. Colleges have football teams in Canada like the United States. Of course, the NHL Montreal Canadians Hockey team reside in Montreal. Montreal seems like Europe at half the price. It is a unique city with the flare of a modern metropolis and the problems of a large city.

Real estate is priced similarly to the United States and US citizens may hold property in Canada. The large numbers of lakes make it ideal for summer fishing trips and outings during the winter. Use caution when driving Canada in the wintertime due to snowy weather and slippery and icy road conditions. Canadians are friendly to Americans even if they do not agree with our international trade policies and they appreciate our business. The homes in Montreal are built to survive the harsh winters. The temperatures in the summer are pleasant without any humidity and the winters are cold similar to northern Europe.

Figure 51. Montreal

61. Tallahassee

Tallahassee is both a state capital and college town with Florida State University and Florida A & M universities near the town center as advisory capacity to the governor. The state capital building is located at the CBD of the city and the federal court is nearby. The architecture is a mixture of old and new buildings with beautiful trees around the downtown area. The city extends 7 miles from the CBD and like other Florida cities allows for suburban development within this boundary. The main highways into the city are off of Interstate 10 which runs next to the city. It's rural setting is both charming and charismatic. There are several golf courses as with most of Florida's major urban areas. The people were very friendly and happy to see tourists which is the primary economic income in Florida. The first time visitor is taken back by the majesty of the surroundings in a climate that is most comfortable at all times during the year. Route 90 is the main highway into the city and major buildings are off this highway which runs east and west.

The downtown area is very similar to smaller college towns in other states where major universities are located. Florida State University was once a women's college in the early 1900's and is home to the ACC Seminoles. The governor of Florida has access to the library and research facilities of the university within walking distance of his executive office. The college is typical of most in America with students congregating at local establishments along the route 90 corridor and on

campus in spaces provided for them near libraries and research facilities.

Figure 52. Tallahassee

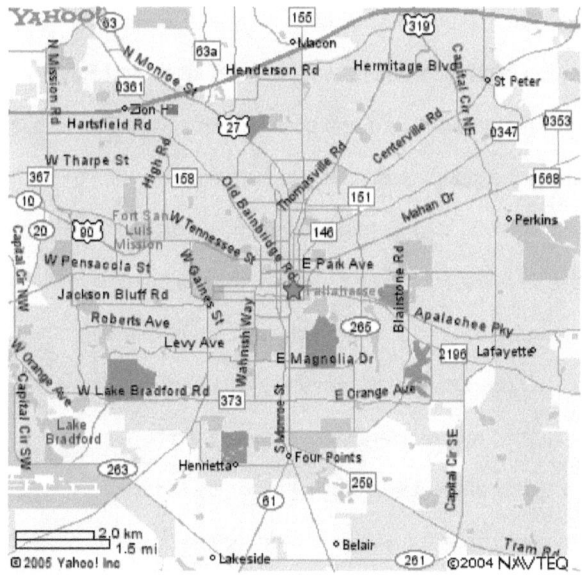

62. Honolulu

Mayor: Jeremy Harris (to Jan. 2005)

2003 Population: 380,149 (46th in US)

Honolulu is a beautiful city on the island of Oahu in Hawaii. It is rich with historical sites and information that helps understand the past kings and Royalty of Hawaii. Diamondhead is an old volcano on one corner of Honolulu. The Zoo is world known. Pearl Harbor is the site of the Japanese surprise attack on the American fleet in the Pacific in World War II. Waikiki beach is a great tourist spot in Honolulu and is worth the time invested to see. Hotels line the beachfront. A drive around the island takes about an hour. The highways are very modern and seeing the island by car is very practical an inexpensive. The cost of living in Hawaii is very expensive. Homes and residences cost more than here on the continent. They import a majority of the items they need for life there. Honolulu is also the government capital of Hawaii. Oahu means "meeting place". Figure 53. Honolulu

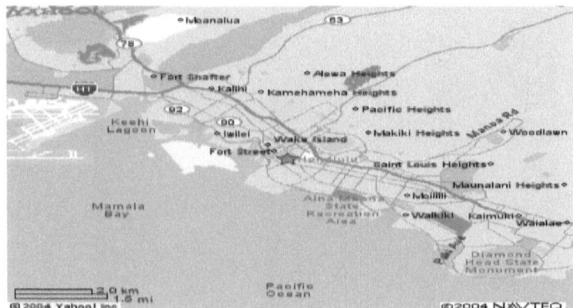

63. Columbia, Maryland

Columbia is a new town in Maryland near Baltimore and Washington. It is formed of villages that each have a shopping center and a town center around a mall with the CBD formed by the businesses located near the mall.

Figure 54. Columbia, Maryland

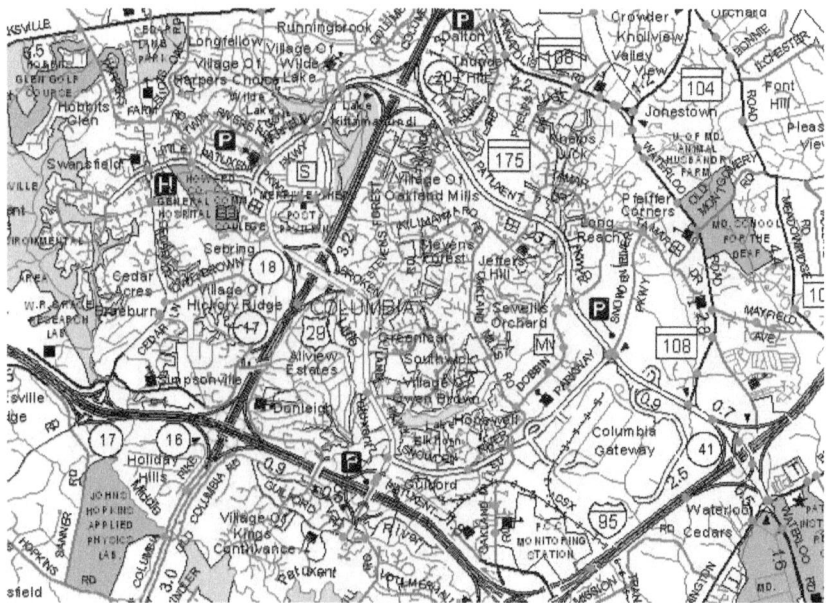

64. Columbus

Mayor: Michael B. Coleman (to Nov. 2007)

2003 population: 728, 432 (15th in US)

    Columbus, Ohio is the capital of Ohio. It is located along I-70 in central

Ohio. It was one of the first cities to have an interstate bypass in the

1960's. The CBD is visible from I-70 when traveling through Columbus to

other destinations. Traveling north from Columbus one can reach

Cleveland. Traveling west one soon comes to Dayton, Ohio on I-70.

Figure 55. Columbus, Ohio

65. Denver

Mayor: John Hikenlooper (to June 30, 2007)

2003 population: 557,478 (26th in US)

Denver is the Mile High city because of her location in the Rocky Mountains. The city is served by Stapleton airport. The downtown area is located just off Interstate Route 25 which runs north and south. The weather is very cold and snowy in the winter and very hot and dry in the summer without humidity. The nature in Colorado makes it a nice place to go on weekend day trips to various naturally beautiful places. The city recently got a major league baseball team. The Denver Broncos are the football team and the Denver Nuggets are the basketball team. Colorado Springs is located 60 miles to the south and Cheyenne, Wyoming is located to the north. Lowery Air Force Base was once a personnel center in downtown Denver and also served as the site of the Air Force Academy when it opened. Some of the public local schools are the University of Denver, Colorado School of Mines, University of Colorado, and Aspen University. The suburbs of Aurora and Arvada are very affluent.

Figure 56. Denver.

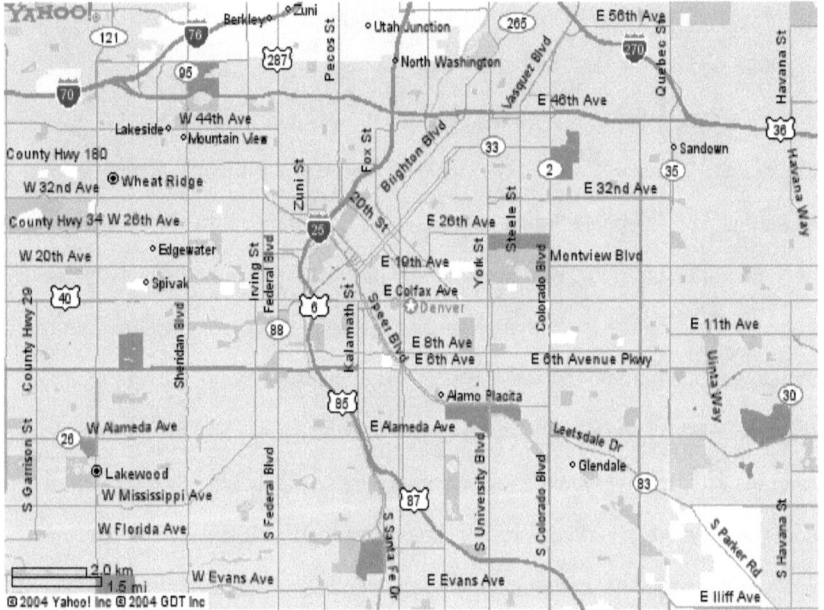

66. Dallas

Mayor: Laura Miller (to May 2007)

City Manager: Teodoro J. Benavides

2003 population: 1,208,318 (9th in US)

Dallas Fort Worth is located in northern Texas. The weather there is hot in the summer and gets cold in the winter with snow from the Rockies. The sometimes also have tornadoes during that season. The city is very affluent and is good for business. Dallas is home of the Cowboys football team and Mavericks basketball teams. Dallas highways have had a high share of accidents and congestion over the years. The CBD is enclosed by a beltway. There are several parks in the downtown area as well as nearby airport as shown in the map below.

Figure 57. Dallas

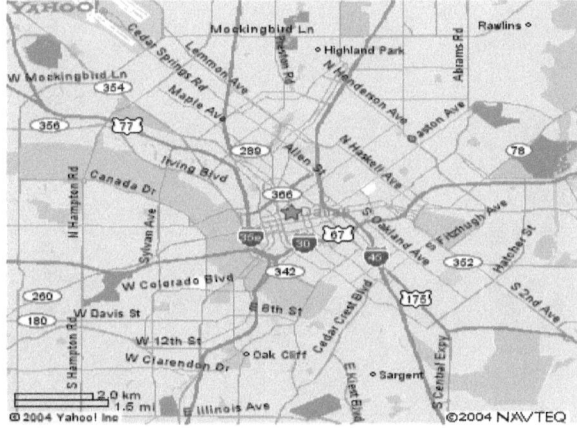

67. Kansas City

Mayor: Kay Barnes (to April 2007)

City Manager: Wayne Cauthen

2003 Population: 442,768 (38th in US)

Kansas City is located on the border of Kansas and Missouri. The name
Kansas come from the Native American tribe who lived there before the white
man. Kansas City is a large stockyard and train depot from the early days of the
west. Texas cattle are shipped north by train to Kansas City and then Chicago.
Kansas City is home the baseball Royals and football Chiefs who play at
Arrowhead Stadium. Kansas is a large farm state with all the produce being
shipped by train through Kansas City to points east. From the map you can see
a river runs through the city. The freeways make travel easier through the city.

Figure 58. Kansas City

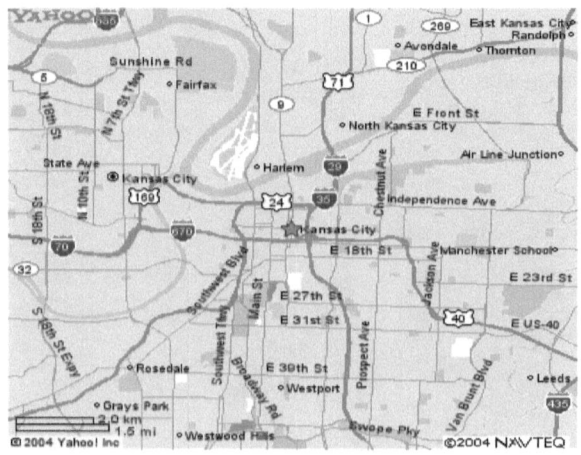

68. San Francisco

Mayor: Gavin Newsom (to Jan. 2008)

2003 Population: 751,682 (14th in US)

San Francisco is named after St. Francis of Assisi. It is one of the only cities named after a Saint. It is also known as the city beside the bay. The international airport is near the bay and one can see Oakland from the terminal. The trolley is famous and still operational on the downtown hills. Fisherman's wharf is also a famous attraction. The Golden Gate bridge is a famous landmark near the city connecting san Francisco to Oakland as one of the largest suspension bridges in the world. San Francisco is home to the Giants baseball team, 49ers football team, and Warriors basketball team.

Figure 59. San Francisco

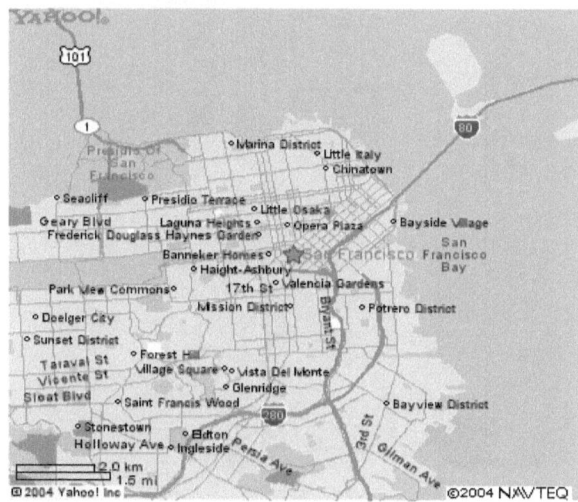

69. Los Angeles

Mayor: James K. Hahn (to June 2005)

2003 Population: 3,819,951 (2nd in US)

Los Angeles is known as the City of Angels. It is well known for Hollywood and acting. The highways are well developed and some of the largest interstates in the country. It is number one in traffic congestion. Los Angeles is the largest of the California cities by population. The climate is very mild year round and this attracts many people. The sports teams are the Dodgers, Angels, Lakers, and Clippers.

Figure 60. Los Angeles

70. Houston

Mayor: Bill White (to Dec. 2005)

2003 Population: 2,009,690 (4th in US)

    Houston is an oil town. NASA Johnson Space Center is also located there. Hobby international airport is the airport serving all of Texas. It is near the gulf coast of Texas and this enjoys sea trade nearby. The downtown is a crosshatch pattern with several major interstates. The major sports teams are the Rockets, Texans, and Astros.

Figure 61. Houston

71. Cleveland

Mayor: Jane Campbell (to Jan. 2006)

2003 Population: 478,403 (35th in US)

Cleveland is located next to Lake Erie and is subject to winter snow from
lake affect. The city is also near interstate 90 which runs along the top of
America from east coast to west coast. The Rock and Roll Hall of Fame is
located in Cleveland. Many businesses are located downtown. There are
also good schools nearby such as Case Western Reserve, Toledo, Cleveland
State, Ohio State, and Ohio University. The airport is Hopkins International
Airport. NASA Ames Research Center is located there. The professional
sports teams are the Browns, Cavaliers, and Indians.

Figure 62. Cleveland

72. San Antonio

Mayor: Ed Gaza (to May 2005)

2003 Population: 1,214,725 (8th in US)

San Antonio started as the Alamo. The old Alamo is located downtown and is a good tourist attraction. The Alamodome is where the Spurs play. There are several military bases located nearby such as Ft. Sam Houston. The downtown area is segmented by the Riverwalk terrace which is a nice location. The food is Texican which includes both Mexican and Texas style hot spicy foods. The people are nice and expect tourists during the summer months. Laredo, Mexico is a short drive away to the south on Interstate 35 for a day of shopping.

Figure 63. San Antonio

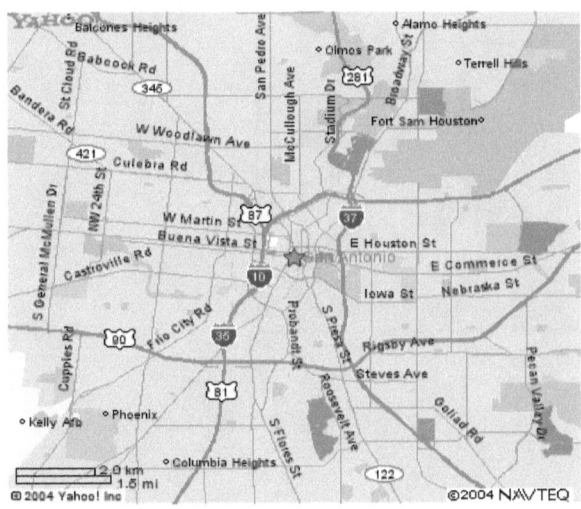

73. Amsterdam

Amsterdam is located in Holland and is an international city as capital of
Holland. It is a city known for it's night life. The downtown area has avenues
which form concentric circles around the city. The train station is located
downtown for easy access to other countries in Europe at inexpensive rates.

Figure 64. Amsterdam

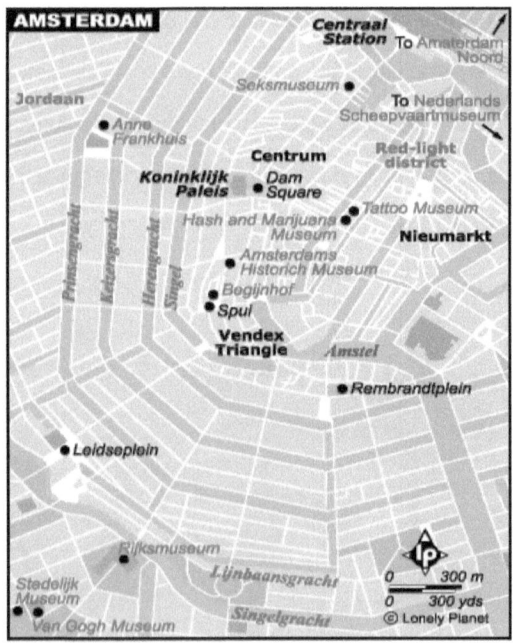

74. Atlanta

Mayor: Shirley Franklin (to Jan. 2006)

2003 Population: 423,019 (41st in US)

Atlanta is gateway to the south. It was home to the Olympics in 1996.
Interstate 75 runs north and south through the city. George Tech University,
Emery, and Spellman are all located near Atlanta. Atlanta is the location of
the underground railroad which helped slaves escape to the north during the
civil war. Rev. Martin Luther King has a memorial in Atlanta. Atlanta was
destroyed during the civil war by General Sherman who marched from Atlanta
to Savannah burning everything on the way. Atlanta is sports home to the
Braves, Falcons, and Hawks pro teams.

Figure 65. Atlanta

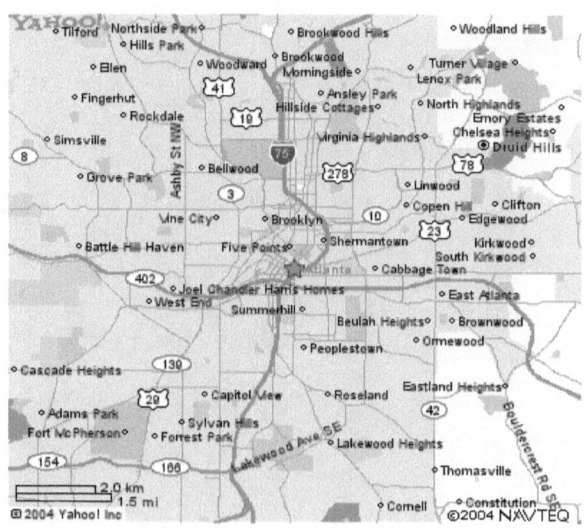

75. Calgary

Calgary is the Denver of the Canadian Rockies. It is located directly north
of Denver but the winters are far more harsh. The winter Olympics were held
in Calgary as a testament to the winter conditions. To the west is open
rugged Canadian country and the Yukon. To the east are the Great Lakes
and the high populated Canadian cities of Toronto, Montreal, and Quebec.
Calgary has the Flames hockey team.

Figure 66. Calgary

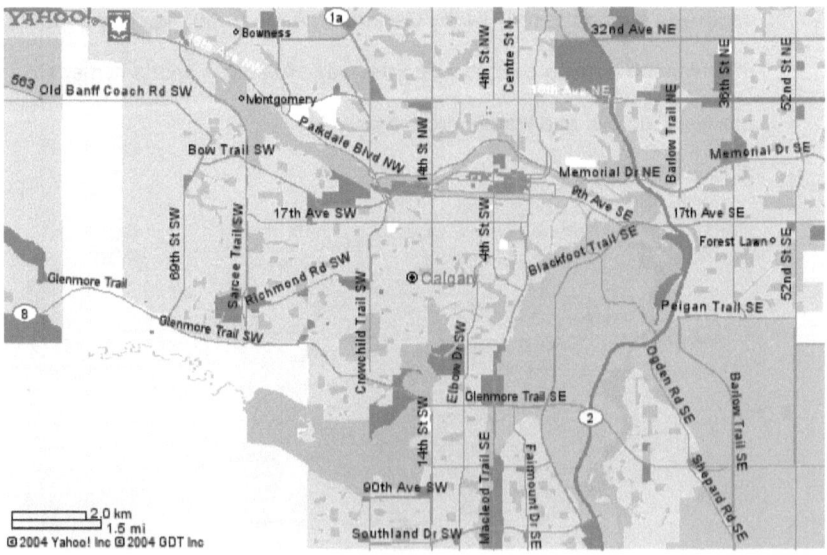

76. Capetown, South Africa

Capetown is in South Africa which was a colonial holding of the British. Apartheid existed in South Africa for many years keeping local black people subservient to the whites. This has now changed with Nelson Mandela as the new president of South Africa and many years of imprisonment from 1963-1982 for political reasons. The street name show the British influence. The train station is located at the center of the city. The streets are in a crosshatch formation.

Figure 67. Capetown

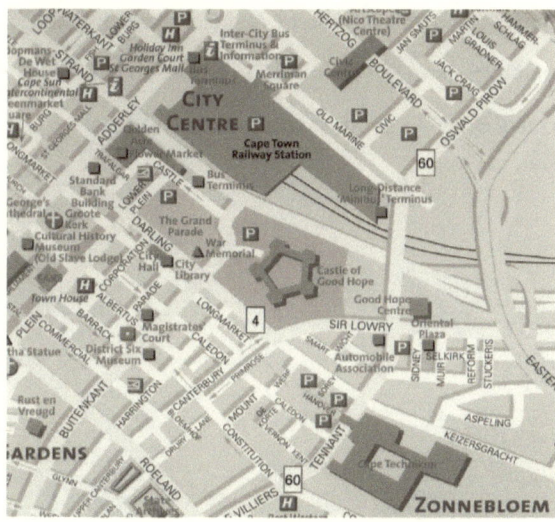

77. Miami

Mayor: Manuel A. Diaz (to Nov. 2005)

City Manager: Joe Arriola

2003 Population: 376,815 (46th in US)

Miami is located in southern Florida. It has beaches and a river next to it.

There are several schools located there such as Miami University. The

weather is the warmest in the states during the entire year which is ideal for

recovery of medical injuries. South of Miami is the everglades national park

system which is sub tropical jungle conditions. Interstate Route 95, US Route

1, and an international Airport connect Miami to the rest of the world. The

downtown area is large and expansive like most Florida cities with city limits

extending far away from the CBD. The sports teams are the Dolphins,

Marlins, and Heat.

Figure 68. Miami

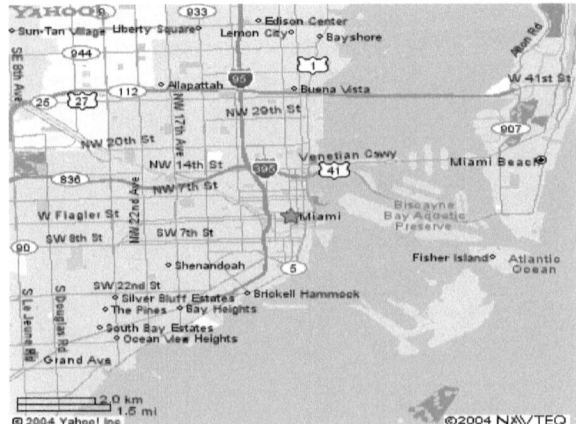

78. Las Vegas

Mayor: Oscar Goodman (to May 2007)

2003 Population: 517,017 (30th in US)

Las Vegas is located in the desert in Nevada. This poses challenges for obtaining water from the Hoover Dam which has to be pumped into the city. This is the city that breaks the rule about being located near a water supply. The city is a center for gambling and adult play. It was also a longtime center of Mafia activities. The city is a day's drive from the west coast and is frequently visited by people from all over the USA. The University of Nevada Las Vegas is located there and is part of the Mountain West division in NCAA athletics. The downtown area is a basic crosshatch design with major interstates running through them.

Figure 69. Las Vegas

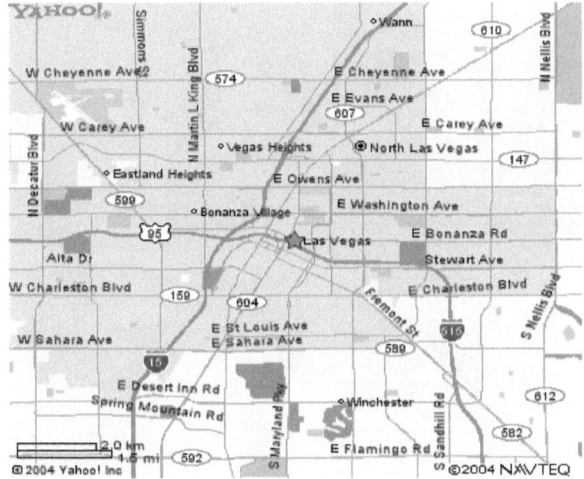

79. Cincinnati

Cincinnati is located in the southwest corner of Ohio. It is one of the largest cities in Ohio. It is on the Ohio river which was once used to ship iron ore to Pittsburgh from other cities in Ohio. Cincinnati is an industrial city in this regard with major factories located there. There are many good universities in the area such as University of Cincinnati. The Reds, Bengals, and old Royals are their sports teams.

Figure 70. Cincinnati

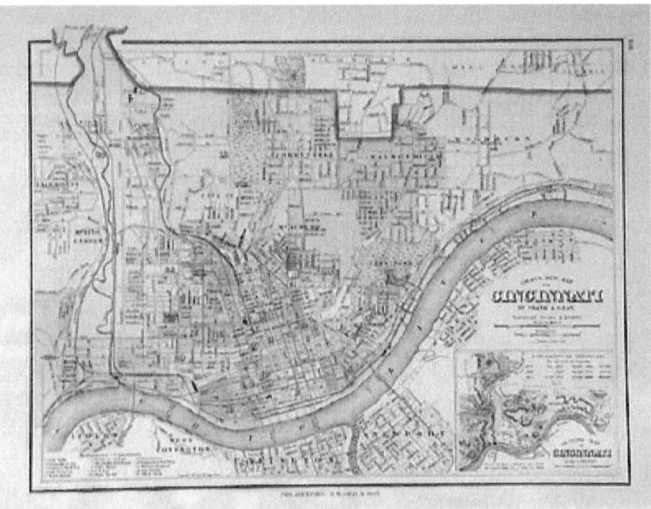

80. Greenbelt, Maryland

Greenbelt is important because it was the federal government who created it as a new town after World War II to house veterans in the 1940's. The old town is a small downtown where walkers are accepted openly and there is one of the first malls in the United States. The city is incorporated as a municipality and a single road runs around the city. Designers of other new towns studied Greenbelt before they built them in Columbia and Reston.

Figure 71. Greenbelt

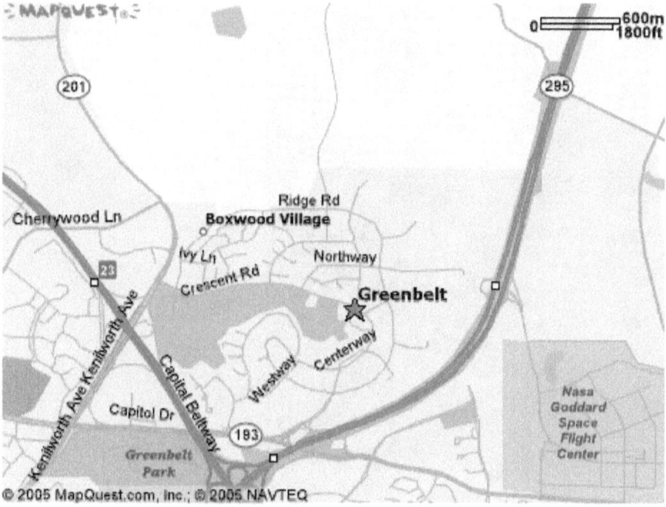

81. Jakarta

Jakarta, Indonesia is a major stopover for many international flights.
It is made of volcanic materials and has a history of eruptions. The island
Was hit by the tsunami of Dec 2004 very hard. The map shows it has a
harbor on the Java Sea and a downtown area.

Figure 72. Jakarta

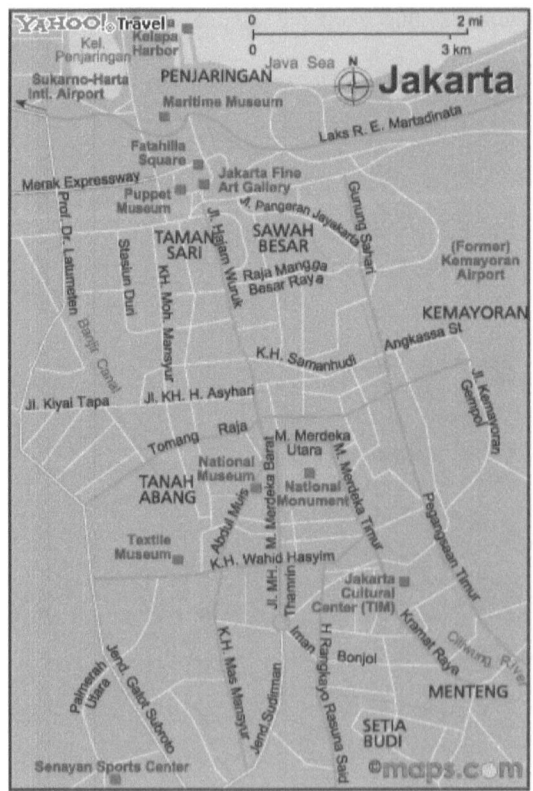

82. New Orleans

Mayor: C. Ray Nagin (to May 2006)

2003 Population: 469,032 (34th in US)

New Orleans is on the Gulf of Mexico and is a major port city. The lowlands around the mouth of the Mississippi River are nearby and form marshlands and the Bayou. Annually, at Easter there is an event called "Mardi Gras" in New Orleans. It is also famous for Bourbon Street and the Redlight district. Since buildings are so low to sea level there are few with basements and cemeteries are specially built above ground in New Orleans. A famous battle in the War of 1812 occurred in New Orleans with General Jackson beating back the British invaders. This was defining moment for the US military.

Figure 73. New Orleans

83. Phoenix

Mayor: Phil Gordon (to Oct. 2007)

2003 Population: 1,388,416 (6th in US)

Phoenix is at the heart of Arizona. The city has grown steadily with people moving from the colder climates in the United States. The suburb of Scottsdale is very affluent and has the most expensive homes. The airport is Sky Harbor International Airport. The sports teams are the Diamondbacks, Suns, and Cardinals. A large contingent of native Americans lives in Arizona. The major universities include Northern Arizona, Arizona State, and the University of Arizona.

Figure 74. Phoenix

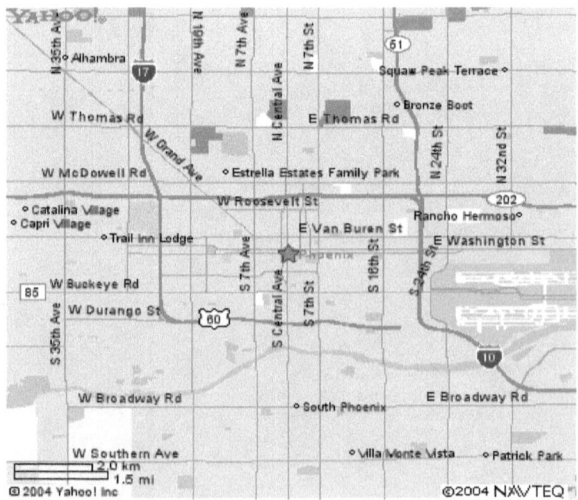

84. Reston, Virginia

Reston is important because it is also a new town built in the 1960's to meet the demands of the new suburban living near a major city (Washington DC). The new town was primarily a bedroom community with only a shopping mall and several businesses located there. Today, the number of businesses there are tremendous as companies have figured out that locating near the population centers is how to get workers involved in the company.

Figure 75. Reston

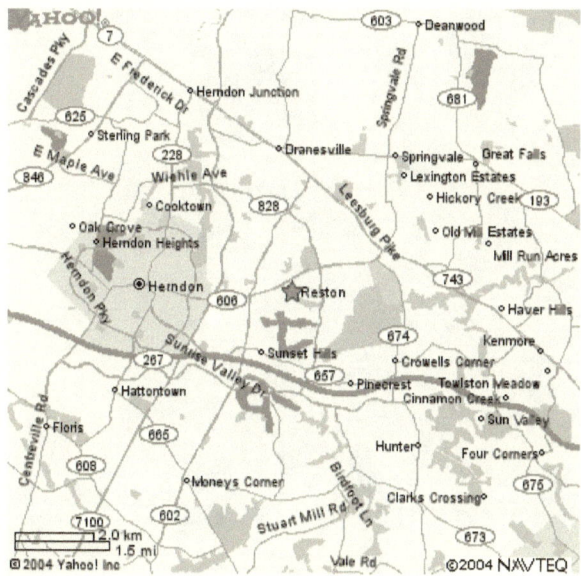

85. Richmond

Richmond was the capital of the confederate states during the civil war. Today, it is a thriving metropolis. The tobacco industry has many factories in Richmond. Richmond has the capitol functions for the state of Virginia. Schools like University of Richmond and Virginia Commonwealth University grace her downtown. Her skyline is impressive when one drive route 95 through her.

Figure 76. Richmond

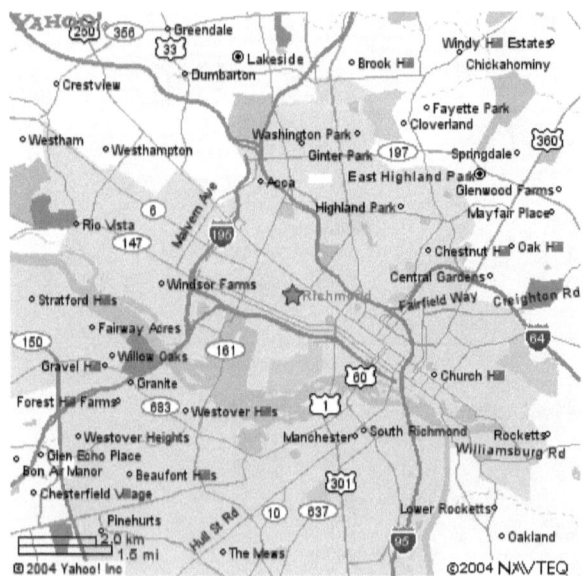

86. Sydney

Sydney is the capital of Australia and located on the northeastern coast of Australia. It is a port city with some beautiful architecture such as the world famous Sydney Opera House. Government buildings are located downtown. The country has it's history in the migration of people from Great Britain as do many other colonial countries such as New Zealand, the United States, Ireland, the Falklands, and India. Since Sydney is in the southern hemisphere, they experience winter we the United States experiences summer and they experience summer when the US experiences winter. A train stations is downtown and many road names are distinctly British in origin as illustrated in figure 77. They have sports teams in soccer, rugby, and cricket.

Figure 77. Sydney

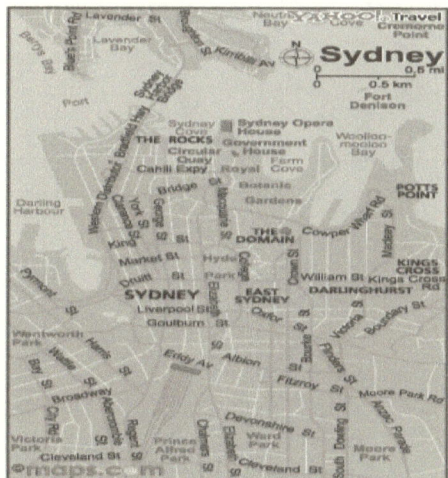

87. Salt Lake City

Salt Lake is home to the Mormon Church of Latter Day Saints. The Lake
is salty rather than freshwater which is why it was given this name. There is a
nearby Air Force Base and both Utah and Brigham Young Universities are
located there. Computer software is one of their city exports.

Figure 78. Salt Lake City

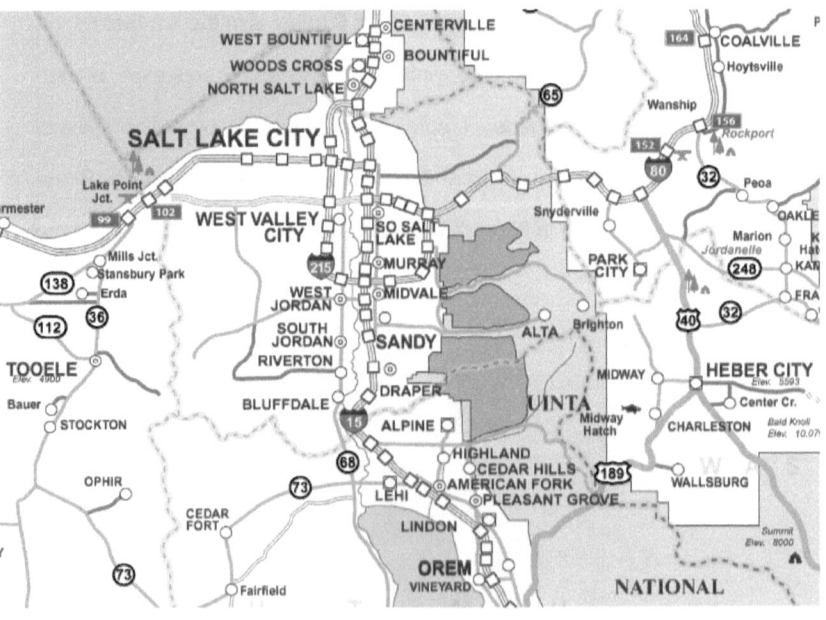

88. Saigon

Saigon is the capital of South Vietnam. It was renamed Ho Chi Mihn City in 1975 after the fall of the city to the North Vietnamese. It was the center of the US Military forces during the Vietnam War. Many people from Vietnam and Saigon were forced to migrate to the United States after the war and live here now under our protection. The city is basically crosshatch roadways with major boulevards overlaid on the smaller roads. The traffic includes many bicycles on a daily basis. In the late 1950's Saigon was held by the French. The language is Vietnamese.

Figure 79. Saigon

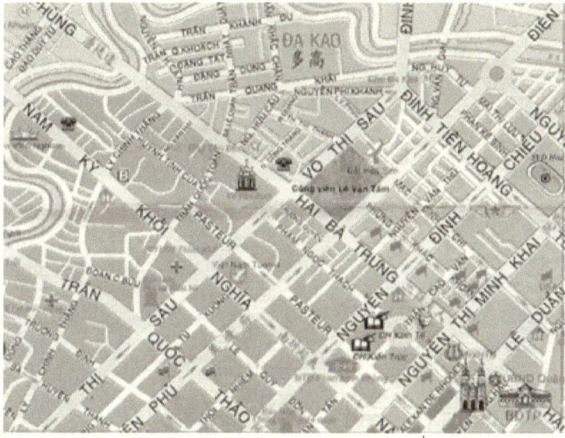

89. Sacramento

Mayor: Heather Fargo

2003 Population: 445,335

Sacramento is the capitol of California. It is located near the 39th parallel in north central California. The downtown is a crosshatch CBD and suburbs are located off the interstates. A river cuts the city in half and provides fro drinking water and sewage treatment facilities.

Figure 80. Sacramento

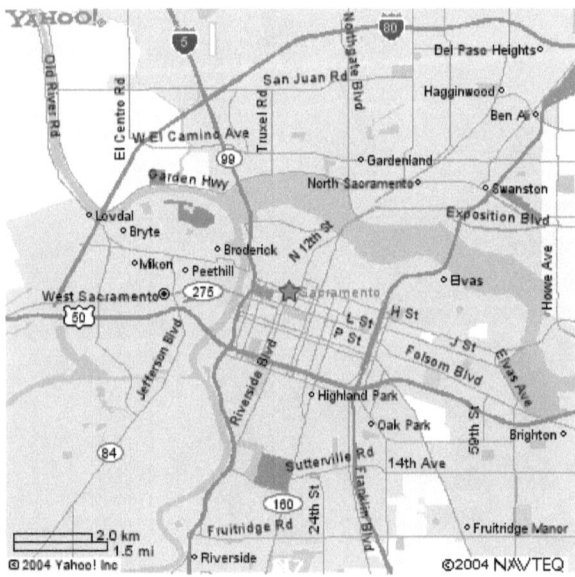

90. Tehran

Tehran is the capital of Iran. The city is located in modern Iran. The society is closely watched by the west for signs of terrorism activity. The government is run by Muslim fundamentalists. They do not have nuclear power at this point in time, except for peaceful purposes.

Figure 81. Tehran

Part III – Future Cities

91. Mars Base Alpha

This city will be one of the first non-earth cities mankind will develop in the solar system. After we landed on the moon in 1969, which I guarantee was a real event, we started building models of space colonies and many people speculated on what these colonies would look like. A wheel shape was assumed or Burnel sphere with neighborhoods and agriculture growing inside the environment.

Figure 82. Exterior Bernal Sphere

Figure 83. Interior Bernal Sphere

Figure 84. Agriculture Bernal Sphere

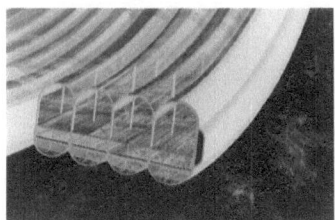

Figure 85. Exterior Teroidal Space Colony

Figure 86. Interior Teroidal Space Colony

A city on Mars would have special requirements that urban planners of earth have not dealt with yet. They would have to ensure a bubble of oxygen around the city of some sort or an enclosure that keeps excess nitrogen and hydrogen outside the city breathing air supply. People living there would most likely have buildings that were also environmental controlled and built by NASA on the surface of Mars. If however, Mars is found to have ancient underground rivers there may a possibility of building beneath the surface. It would be expensive but it would also last the test of time against any Martian weather that might harm a surface dwelling society. Mars Base Alpha would be man's first space city and he could expand to other planets from there. This would make it a landing stop for freighters and mining vehicles bringing back minerals from the other planets to earth. Man will eventually stop his look for extraterrestrials and become one himself by colonizing near and far space. This is when we will make discoveries that will surpass the limited knowledge we currently have on earth. Mars Base Alpha will have rovers that will be powered by solar and other means rather than fossil fuels. The homes will be self sustaining for years without re-supply from earth. Mars gardens will be grown in artificial light and vegetables will help sustain the inhabitants of Mars Base Alpha.

The city will have all the necessities and technology of home except outdoors play will not be permitted. The schools will be teaching space sciences as well as religion and the other basics we teach on earth. The neighborhoods will be connected by either underground walkways or by above ground tunnels. The sun will warm the planet and provide most of the energy needs. The minerals in the planet will be researched and new discoveries and uses for them will be found. A refueling dock on a space station above the planet will guide intra-planetary voyages from one planet to another. We have assured this with our own building of the space station here on earth. A super shuttle craft will be developed with the range to traverse the miles in shorter time frames than we currently can. The current shuttle will become obsolete or only used for above earth travel. There may be some privatization of space as new companies form to compete for contracts and business on Mars Base Alpha[19]. NASA may be taking the orders for various items now. The President has challenged us to go to Mars and we should fully embrace this opportunity. Ex-Astronaut Dr. Buzz Aldrin never received the publicity that his friend Senator John Glenn received, but he had this vision in 1999 when he shared it with a crowd of students at American University. He even showed the students what the launch vehicles would look like. Many of them were not even alive when he landed on the moon. He wrote a book on the topic of privatized space travel and believed that man would make it to other planets this way.

---

[19] Mars and Return to Mars, Science Fiction Novels by Ben Bova.

92. Epilogue

Urban planning is an ideal situation where we can influence how our local communities will look in the future. It takes legislators, planners, engineers, and private citizens working together to make the whole process happen effectively. We do not have the luxuries of creating an entirely new world except on other planets where we are doing urban colonization of outer space which is a whole separate discussion requiring much more resources. The environments we create to live in on earth are basically in harmony with the surroundings and with the climates we face. We can look at all the major cities of the world and provide guidelines that describe the patterns of development and planning components but only when we live with our decisions can we see the long term impact of the choices we have to make. For example, no one would chose to live in a slum or ghetto if they had better economic means to live a higher quality life. The bottom line is improving the quality of life for all our world citizens and improving the human condition, in general in the process. This would describe ideal urban planning that encompasses every member of society, not just the rich and upper classes but also the poor and poverty stricken. It is hoped that you have enjoyed this book and that it has enlightened your views of urban planning at the regional state and local level. If it has helped one individual understand the processes and concepts better then it has been a success. Keep on building a new future.

Bibliography

Aldrin, Buzz, Dr., Public Speech at American University, Washington DC, February 1999.

ASCE (American Society of Civil Engineers) Magazine Series, 2005.

Brunner, Borgna ed., Time Almanac 2005, Information Please Pearson Education Company, 2004, Needham, MA.

Bova, Ben, Return to Mars, Random House, New York, 2003.

Gans, Herbert, The Urban Villagers, New York, 1968.

Gist, Noel & Fava, Sylvia, Urban Society, 6th Edition, Crowell Company, New York, 1974.

Herman, Arthur, How the Scots Invented the Modern World, New York, 2003.

Howard County Office of Planning and Zoning, Howard County Master Plan 2000.

Howard County Office of Planning and Zoning, Ellicott City Master Plan, 2004.

ICMA, Principles and Practices of Urban Planning, Washington DC., 1968.

ITE (Institute for Transportation Engineers) Journal Series, 2005.

Maryland Planning Office, Maryland Master Plan 6 Year Review, 2004.

McKeever, J. Ross, The Community Builders Handbook, Anniversary Edition, Urban Land Institute, Washington DC, 1968.

Meyer Micheal D., Urban Transportation Planning: A Decision Oriented Approach, McGraw Hill, New York, 1984.

Moynihan, Patrick, Senator, The Melting Pot, New York, 1974.

Northam, Ray M., Urban Geography, John Wiley and Sons, New York, 1975.

Robert Moses: Powerbroker

Schuman, Howard, Building the American City: Report of the National Commission on Urban Problems to the Congress and to the President of the United States, House Document No. 91-34, GPO, Washington DC, 1968.

Websites

www.ho.co.gov

www.mpo.gov

www.mec.gov

www.sha.state.md.us

www.whitehouse.gov

www.umd.edu

www.hcc.edu

www.yahoo.com

www.google.com

www.apa.org

www.jhu.edu/apl

www.umd.edu

www.nasa.gov

www.hud.gov

Biography

Donald Joseph Gray Chiarella was born in Kilmarnock, Scotland in 1956. He lived in Topeka, Kansas at Forbes AFB until he was 10 years old. He has lived in suburban Maryland since 1966 and attended local high schools and colleges and taught at several of them. He served in the Air Force one year and as a US Navy civil servant Computer Specialist at Bethesda Naval Medical Hospital for 10 years. He then worked for 5 government contractors as a project leader in the Washington DC metropolitan area. He was commissioned a GSA executive agent in the Governmentwide IT Policy Office for seven years. Then he worked for two contractors at a bank in Delaware and the Marine Corps at Quantico on the SABRS Accounting System project. During this time he taught at UMUC and AACC colleges. He has been working as a Supervisory DBA and MIS/IT Section Chief for 8 years at Maryland State Highways managing the MAARS Statewide Accident Database, Applications Development, and Statistics production for politicians and other government and private customers. He has coached little leagues through high school basketball and baseball. He was a league commissioner. He was a PTA Delegate to Howard County. He is a Howard County Democratic Chief Judge. He is a presidential scholar. He was a state merit scholar and NHS member. He was Vice President of DPMA American University Chapter. He was president of the United Methodist Men twice and lobbied Congress in 1993 for World Peace. He is a past member of DPMA, New York Academy of Sciences, PMA, NARFE, University of Maryland

Alumni Association. He is currently an active member of IEEE, MAA, ITE, NAS-TRB, USAF Academy AOG, US Naval Institute, ACM, American University Alumni Association, Maryland TRCC, and Elkridge PTA. He holds a BA in Urban Planning and IFSM from University of Maryland (1979) and a MSTM degree in Technology Management from American University (1988). He earned his Ph.D. from Kennedy Western University in Management Information Systems (2001) and was one of the first online Ph.D. recipients. He is certified in Federal Government Contract Management by George Washington University (1996). He is a certified Computer Security Manager (2004) by ISACA, certified Urban Planner (2004) by Univ. Maryland, and Certified Data Resources Management Professional (2004) by the ICCP in Chicago.

He has published 6 books and over 100 articles and plans for the government. He has also built more than 20 computer software systems including major operating systems, databases, and telecom systems. He has helped direct funds to major government projects such as USCG Academy Telecom System, NASA projects, and other major government contracts.

He currently lives with his wife Mimi, a master teacher in Howard County, and 4 great children (Donald, Mia, David, and Michaela) in Hanover, Maryland. He loves touring cities and historic sites on vacations, building models, gardening, and sports.